Techniques of Nucleic Acid Fractionation

S. R. Ayad

Department of Biological Chemistry, University of Manchester

With the assistance of

J. Blamire

Post Doctoral Research Fellow

WILEY—INTERSCIENCE

a division of John Wiley & Sons Ltd

LONDON NEW YORK SYDNEY TORONTO

Library of Congress Catalog card No. 78-39232

ISBN 0 471 03846 6

Printed in Great Britain by
Dawson & Goodall Ltd.,
The Mendip Press, Bath

This book
is
dedicated
with warmest affection and admiration
to
G. R. BARKER

Preface

One of the most important aims of the scientist is to subdivide the object of his investigation into a large number of small components which are themselves complete, indivisible and discrete entities and can be analysed individually. The subsequent, and more difficult, aim is to reassemble all the parts in such a way that the degree of co-operativity between the various parts, allows the completely integrated object to function in the correct way. The physicist was not satisfied until he had 'smashed the atom' and studied the subatomic particles. In the same way, the biochemist will not be content until he understands how the various components of the cell work in conjunction with each other to produce 'life', that is, until he has synthesized a living cell. Contrary to popular belief, the creation of life is not as imminent as would first appear: the biochemist does not know the mechanism of many cellular reactions. Indeed he has not yet isolated and sequence analysed a complete chromosome, which holds the knowledge for every functioning part of the cell.

The biochemist analyses the cell by isolating the subcellular particles and the various chemically distinct macromolecular groups (nucleic acids, proteins, polysaccharides and lipids) contained within and outside (cell cytoplasm) the particles. Since within each class of macromolecules, there is a whole spectrum of chemically distinct molecules, which consist of a varying number of the same building blocks assembled in both a different order and conformation, the problem of isolating intact components which are no longer divisible (except by degradation), becomes a formidable one, and relies on elegant fractionation techniques.

This book attempts to collate some of the main techniques applied in the fractionation of perhaps (and I say this 'with tongue in cheek') the most important class of macromolecules—the nucleic acids. It is intended to provide theoretical as well as practical knowledge for third year undergraduates, and to help those who are carrying out research at the postgraduate and higher levels to choose a particular technique, or combination of techniques, best suited to their particular problem. The choice of technique(s) will depend on many factors, firstly the type of purit of nucleic acid species (DNA or RNA), secondly, the quantity of starting material available, thirdly, the speed of fractionation, and finally, the stability of the macromolecular species during fractionation. The merits and disadvantages of the techniques will be compared and contrasted by the use of specific examples, many of which involve nucleic acid metabolism rather than structure. For this reason, the background literature involved in the more essential experiments has been described fairly concisely with a particular experiment. The first part of the book describes the biochemical and biophysical properties of nucleic acids, because it is these properties on which a particular method depends. The second part is devoted to the fractionation techniques *per se*, together with their various biological applications, the examples of which have been chosen to illustrate a particular technique.

While it will be clear that many problems have been solved in recent years using the techniques to be described, many problems remain unsolved, and will necessitate the formulation of new methods of fractionation. For example, the molecular size of DNA of higher organisms is still uncertain. If one is able to isolate a whole chromosome intact in an appreciable quantity, then the problem of sequence analysis of such a large molecule becomes formidable. Messenger RNA is probably transcribed complementary to only one DNA strand. Which of the two strands, and why, has still to be determined, and will require specific hybridization techniques in addition to specific fractionation techniques. Also, the isolation and fractionation of a single messenger RNA which will code for a specific protein has still to be determined. A further problem is that of the mechanism of mutagenesis. It will be important to isolate the specific segments of DNA which contain the mutated genes and hence lead to defective proteins. These are just a few of the many problems facing the nucleic acid biochemist.

There are many without whose help and encouragement this book would not have been written. I shall always be indebted to Professor G. R. Barker, who first made it possible for me to undertake nucleic acid research, and who during the past eight years has helped me in so many ways, and become both a valuable colleague and friend. There are insufficient words to express my gratitude to my wife, Shirley, who has constantly encouraged me, and helped me by reading each ensuing draft of the manuscript. I thank her also for enduring the sight of me thinking and writing night after night, and for allowing chaos to reign in the home. I should also like to thank my colleagues, Drs. W. D. Stein and C. H. Wynn for many stimulating discussions and for their friendship and help over the years. I wish to thank Miss Margaret Barber for her excellent typing of the preliminary drafts as well as the final draft of the manuscript, and Mrs. M. Cubbon for helping to type part of the manuscript. I should like also to thank my technical assistant, Mrs. I. Dimond. I gratefully acknowledge permission from many authors and publishers to produce their diagrams from scientific journals. Finally I should like to thank very much my publishers, John Wiley & Sons.

September, 1971 S.R.A.

Contents

CHAPTER 1

A. The structure of nucleic acids

B. Biological assays of nucleic acids

A. The structure of nucleic acids

Introduction

There are many different types of nucleic acids, each having a unique chemical composition, size, conformation and metabolic role. Each species can be isolated by the fractionation techniques described in subsequent chapters, and its function studied further *in vitro*. Since fractionation occurs by selectively removing the various components by virtue of their different chemical and physical properties, knowledge of these properties is required to choose a particular fractionating technique. The physicochemical structure of all the known nucleic acid species therefore is described briefly in this section.

Polynucleotides

The nucleic acids are usually considered under two broad categories: deoxyribonucleic acid and ribonucleic acid. Both are macromolecules of extremely high molecular weight and are negatively charged at physiological pH. They contain about 15 per cent nitrogen and 10 per cent phosphorus, possess one gram molecule of a sugar (D-ribose in RNA or 2-deoxy-D-ribose in DNA) per gram atom of phosphorus, and strongly absorb ultraviolet light in the region of 260 mμ because of the presence of the heterocyclic pyrimidine and purine bases. In DNA, the pyrimidines are cytosine (C) and thymine (T) and the purines adenine (A) and guanine (G), while in RNA uracil (U) is present but thymine absent (Figure 1.1).

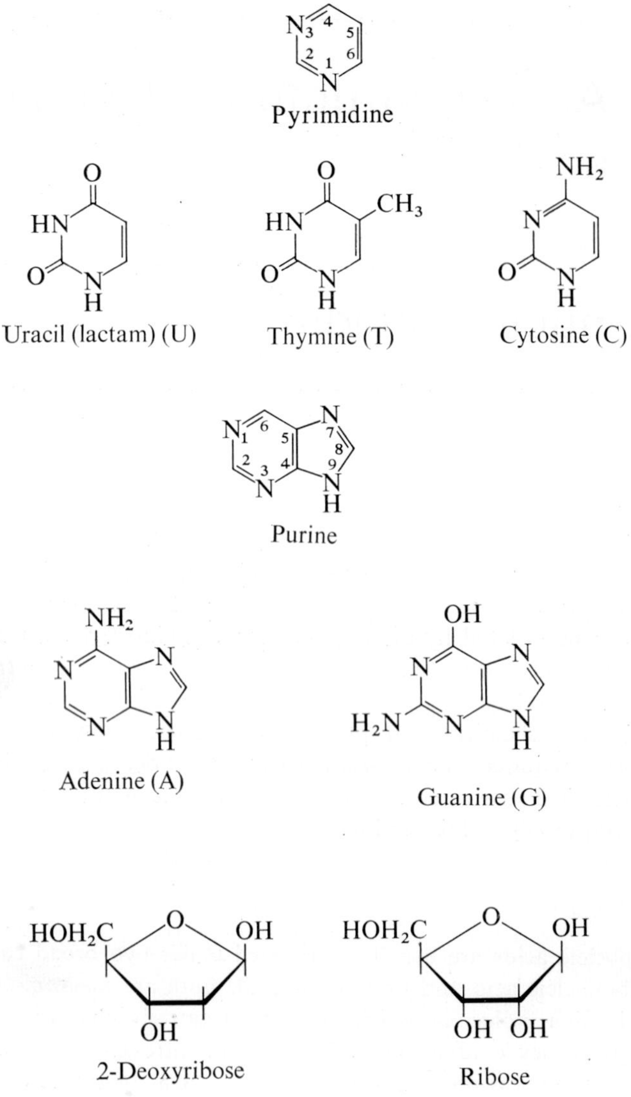

Pyrimidine

Uracil (lactam) (U)

Thymine (T)

Cytosine (C)

Purine

Adenine (A)

Guanine (G)

2-Deoxyribose

Ribose

Figure 1.1. The chemical components of nucleic acids

In addition, other bases occur less frequently in certain types of nucleic acids. They occur rarely in DNA but 5-hydroxycytosine is found in the place of cytosine in T-even bacteriophage DNA. In some DNA's, 5-methylcytosine and 6-methylaminopurine are found in small quantities (Figure 1.2a). Unusual bases occur more frequently in RNA; these include 6-methylaminopurine, 5-methylcytosine, 1-methyladenine, 2-methyladenine, 6,6-dimethyladenine, 1-methylguanine, 2-methylamino 6-hydroxypurine and 2,2-dimethyl-6-hydroxypurine (Figure 1.2b).

These components are assembled into the long-chain macromolecules in a repeating sequence which is similar for both RNA and DNA. A backbone is formed of sugars linked with phosphate bridges, and a heterocyclic base is attached to each of the sugars as shown in Figure 1.3. The chains can be hydrolysed selectively to give nucleosides, nucleotides, or the bases, sugar and phosphate residues in a free form. The nucleosides are the *N*-glycosides of each heterocyclic base. The sugar, which is either D-ribose or 2-deoxy-D-ribose, forms a β-linkage from $N_{(1)}$ of the pyrimidines and $N_{(9)}$ of the purines, and is in the furanose configuration. Thus the correct name for deoxyguanosine would be 9-β-D-2-deoxyribofuranosyl guanine, a name which is nearly always abbreviated usually to dG. Nucleotides are the sugar-*O*-phosphate esters of the nucleosides. Mild hydrolysis of RNA yields a mixture of 2′-, 3′- and 5′-mononucleotides, whereas in the case of DNA only the 3′- and 5′-compounds can be found (the prime indicating which hydroxyl group is esterified on the sugar moiety) (Figure 1.4).

Enzymic and chemical degradation has shown a 3′–5′ phosphate bridge between successive sugars. The primary structure of a single polynucleotide chain is illustrated in Figure 1.5. The base sequences of all but one polynucleotide are unknown, largely for the lack of the appropriate (analytical) methods. The problem, even with the correct methods, is far from simple, as the smallest known single-stranded DNA (the chromosome of the bacteriophage Ø–X174) consists of a circular array of about 5,500 nucleotides.

Deoxyribonucleic acid

The basic structure of the polynucleotide chain has been described but, with the exception of phage Ø–X174 DNA, all the DNA's isolated so far have been double-stranded. The ratios of the constitutent bases: A/T and G/C, in all species, is equal to unity, but the ratio (A+T)/(G+C) varies widely, particularly in the case of microorganisms. Figure 1.4 shows that single polynucleotide chains have a definite polarity determined by the direction of the 3′–5′ linkage, and in the DNA molecule two

(a)

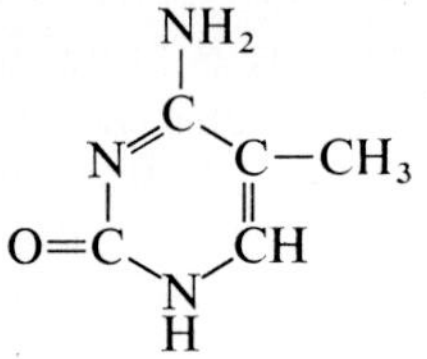

5-Methylcytosine

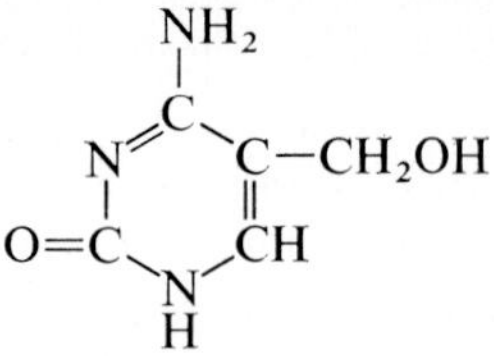

5-Hydroxymethylcytosine

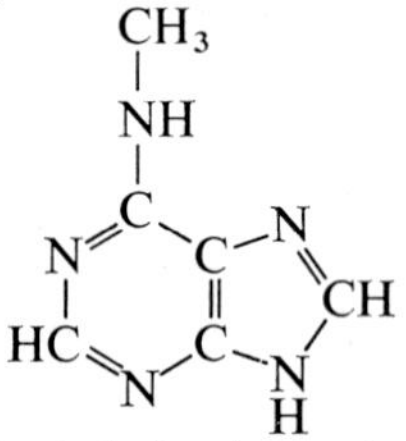

6-Methylaminopurine

(b)

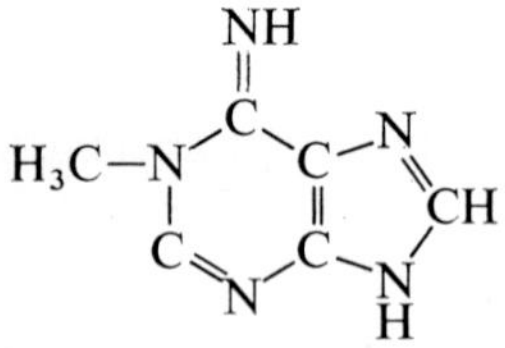

1-Methyladenine

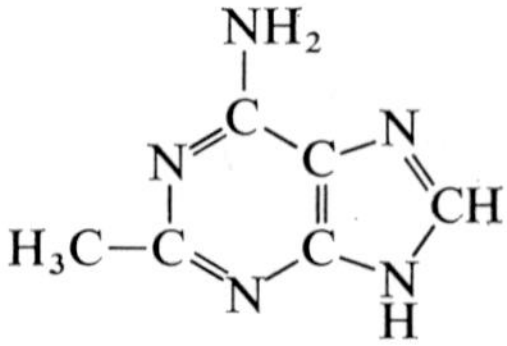

2-Methyladenine

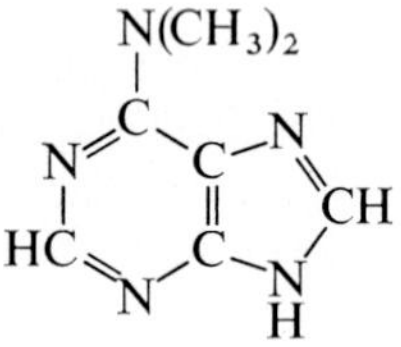

6,6-Dimethylaminopurine

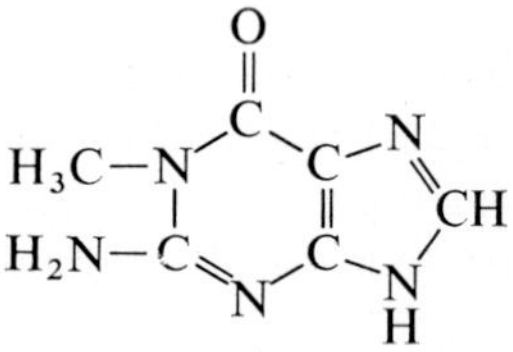

1-Methylguanine

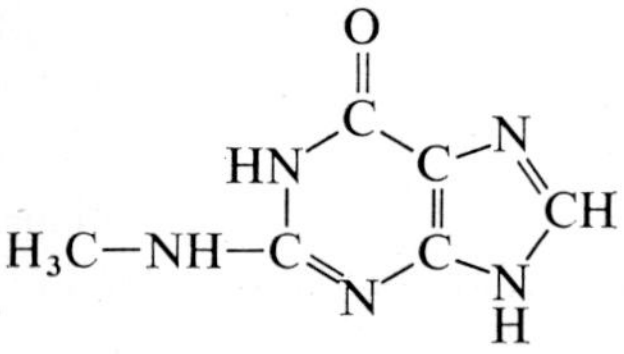

2-Methylamino-6-hydroxypurine

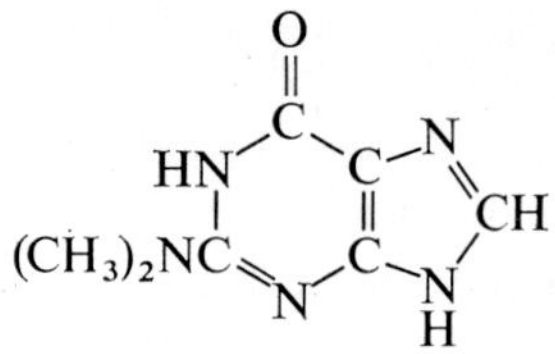

2,2-Dimethylamino-6-hydroxypurine

Figure 1.2. The unusual bases which occur in (a) DNA and (b) RNA

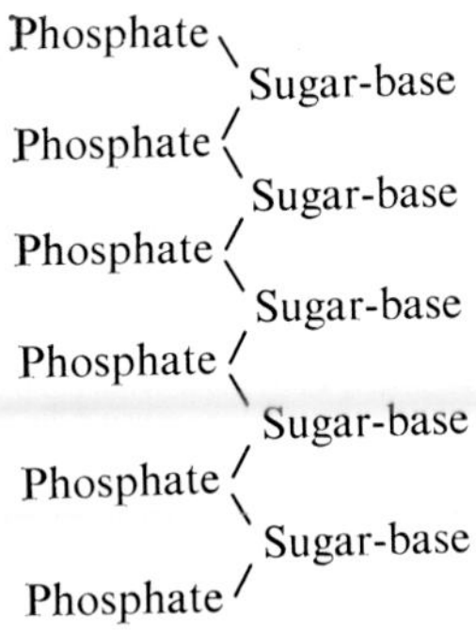

Phosphate
Sugar-base
Phosphate
Sugar-base
Phosphate
Sugar-base
Phosphate
Sugar-base
Phosphate
Sugar-base
Phosphate

Phosphate–Sugar backbone

Polyribonucleotides

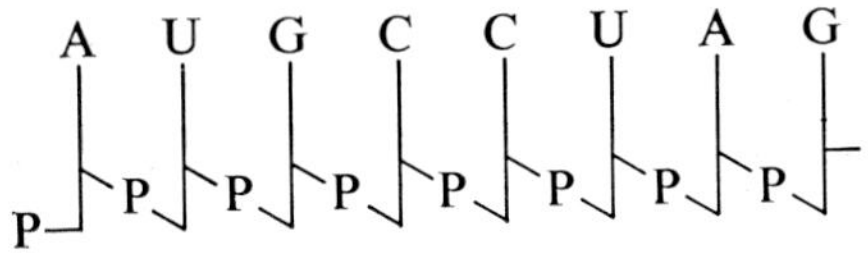

A U G C C U A G

P P P P P P P P

pApUpGpCpCpUpApG

Polydeoxyribonucleotides

dG dC dT dT dA dC dG dA dC

P P P P P P P P P

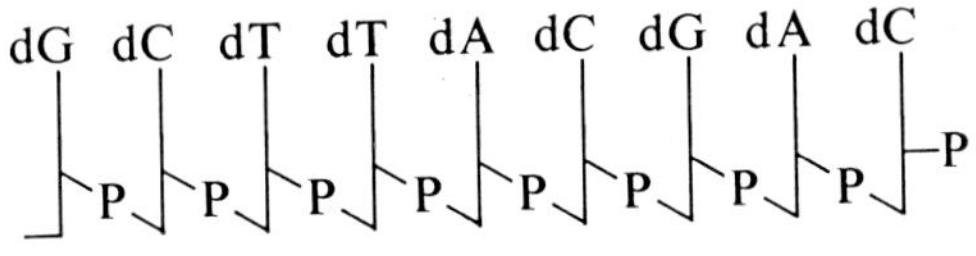

dGpdCpdTpdTpdApdCpdGpdApdCp

Figure 1.3. The structure of polyribonucleotides and polydeoxyribonucleotides

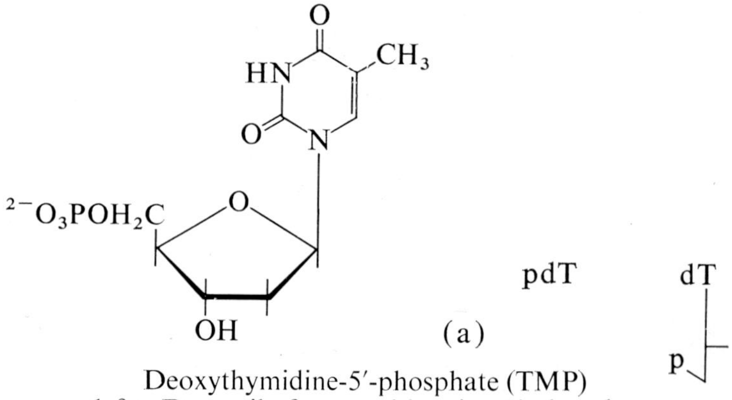

O
HN CH$_3$
O N
$^{2-}O_3POH_2C$ O
OH
pdT dT
p
(a)

Deoxythymidine-5′-phosphate (TMP)
1-β-D-Deoxyribofuranosylthymine-5′-phosphate

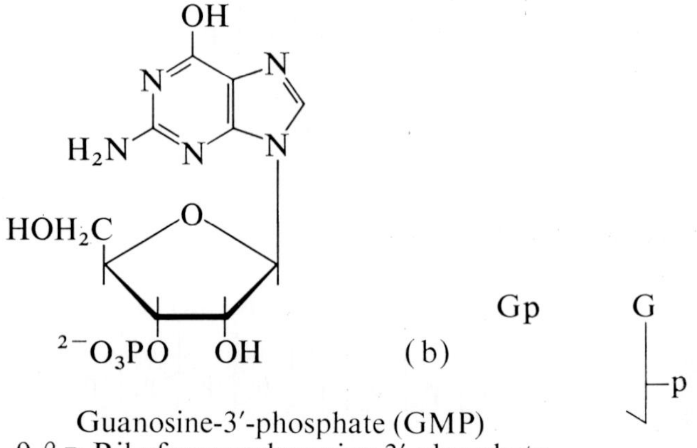

OH
N N
H_2N N N
HOH_2C O
$^{2-}O_3PO$ OH
Gp G
p
(b)

Guanosine-3′-phosphate (GMP)
9-β-D-Ribofuranosylguanine-3′-phosphate

Figure 1.4. (a) 5′-nucleotide of thymine and (b) 3′-nucleotide of guanine

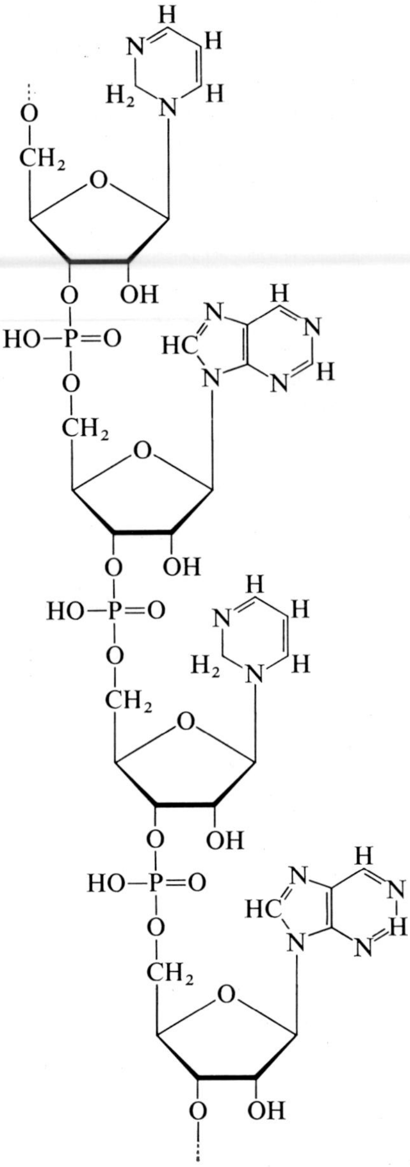

Figure 1.5. Schematic representation of a polynucleotide chain

such chains coil around each other in opposite directions. The effect of this is to produce a double helix, which has the appearance of a spiral staircase, and an outline that shows one deep groove and one shallow groove. The bases, which are largely hydrophobic, stack in the centre of the molecule, whereas the negatively charged phosphate groups line the outer surface which is in contact with the surrounding medium (Figures 1.6 and 1.7). The bases on one chain align opposite to the bases

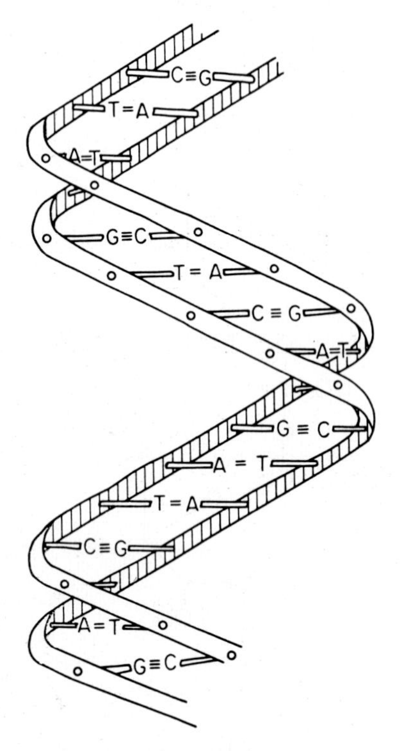

Figure 1.6. Schematic representation of the DNA helix showing the phosphate sugar ribbons on the outside and paired bases bridging the gap between the two strands

on the adjacent chain in such a way that A is always found opposite to T, and G is always opposite C. This accounts for the unity of the A/T and G/C ratios, and implies that the order of bases on one chain determines the order of bases on the other chain. The two opposite bases are held together by a system of hydrogen bonds that not only helps to hold the two

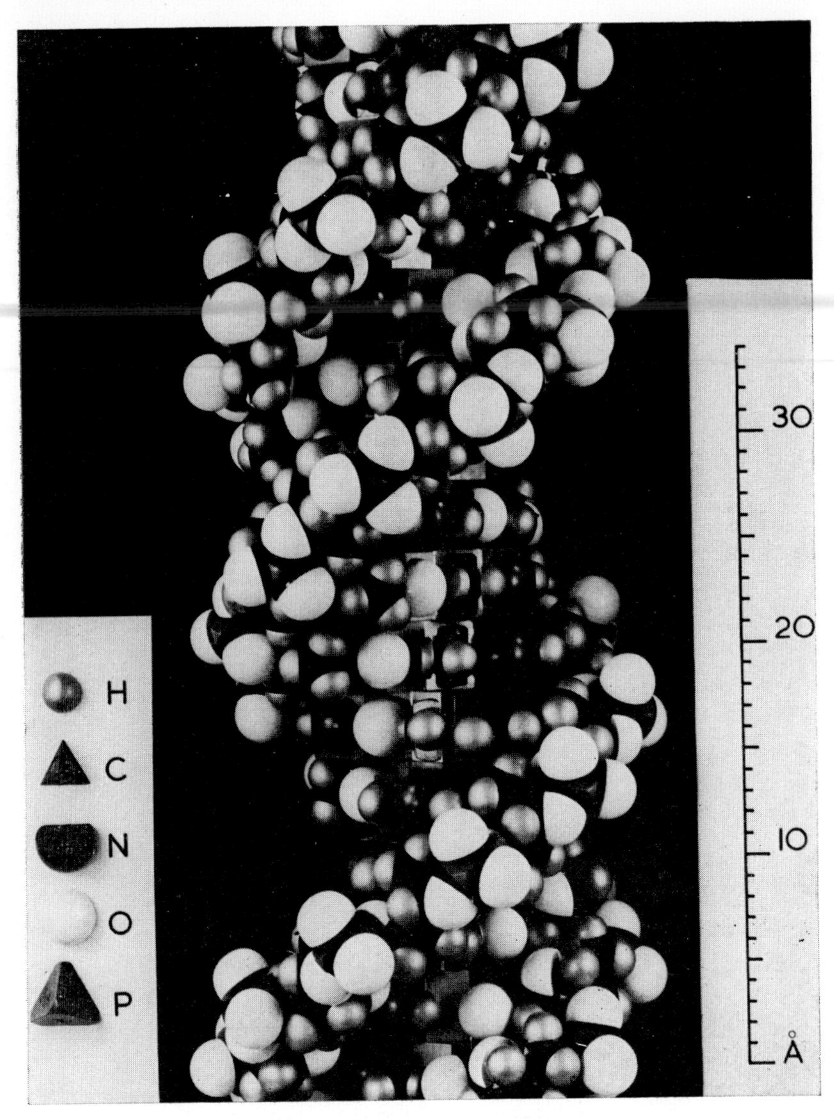

Figure 1.7. A space filling model of a DNA molecule showing the arrangement of various atoms. (By kind permission of Professor M. F. H. Wilkins)

chains together, but also ensures the correct base-pairing properties (Figure 1.8). The bases lie with the planes of their rings approximately at right angles to the helical axis of the molecule, and there are ten base-pairs for each complete turn of the helix. If water is removed from DNA

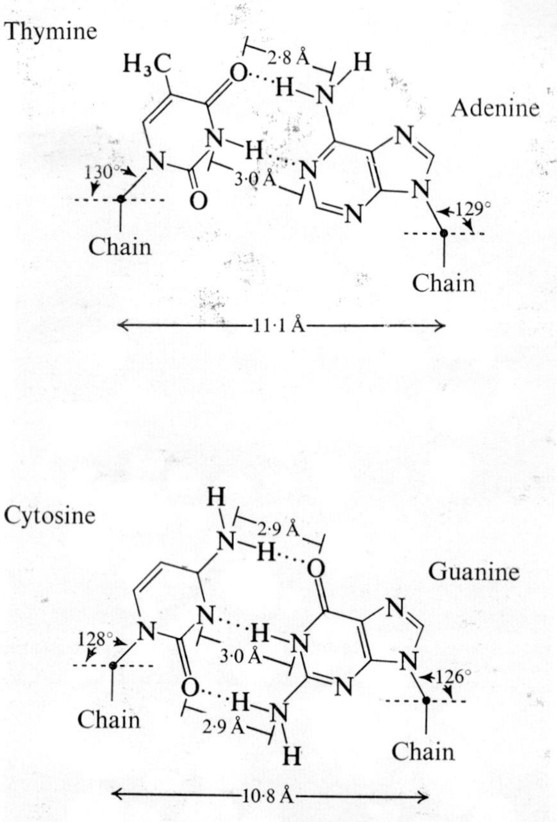

Figure 1.8. The hydrogen-bonding properties of adenine–thymine and guanine–cytosine base-pairs according to the Watson and Crick hypothesis

fibres, the molecules undergo certain structural alterations. The molecule contracts lengthwise, and takes up a new conformation in which there are eleven base-pairs per turn, each tilted at an angle of 20° relative to the molecular axis. The long rigid nature of the DNA molecule makes it very susceptible to hydrodynamic shear forces that degrade the molecule.

In vivo, DNA is rarely found as a completely free molecule. In higher animals, it is protected by specialized proteins usually of a very basic nature, called histones, whereas in bacteria the proteins are replaced by oligoamines such as putrescine or spermine. Viruses present a special situation: the DNA is in an inactive form which is strongly folded and encased in a protein coat. When the DNA becomes metabolically active, the protecting substances are released and/or replaced, allowing the DNA molecules to undergo their metabolic functions (for further details see Hayes (1968)).

In the cells of higher animals and plants, the cellular DNA is, for the most part, concentrated into a specialized body termed the nucleus. Extranuclear DNA has been found in chloroplasts, mitochondria and other potentially self-reproducing organelles, but by far the largest amount is located in the nucleus. The total amount of DNA per nucleus is a very variable quantity, but the average amount is approximately 4 to 8×10^{-12}g., the number of nucleotide pairs in a single human germ cell being of the order of 10^9. Bacterial cells contain about a thousandth of this amount, i.e. 10^6 nucleotide pairs, corresponding to about 4×10^{-3} picogram of DNA.

As in the case of higher organisms, the DNA of bacteria is also not distributed evenly throughout the cell, but is usually concentrated in nuclear regions. They have no nuclear membrane (Ryter, 1968) and no specialized histones. The bacterial chromosome has been shown by autoradiography (see Chapter 2) to consist of a single DNA molecule with a molecular weight of about $2{\cdot}8 \times 10^9$ daltons in the case of *Escherichia coli* (Bleeken, Strohbach and Sarfert, 1966), which is also circular (Cairns, 1963) as shown in Figure 1.9.

Numerous attempts have been made to isolate bacterial DNA with the conventional biochemical techniques and to determine its molecular weight. During these studies it became evident that chain-length varied with the procedure used, and that the apparent molecular weight of DNA could be significantly decreased by mechanical shearing. A molecular weight of 2 to $2{\cdot}5 \times 10^8$ daltons was the highest found with bulk isolation techniques (Kelly, 1967; Klesius and Schuhardt, 1968). However, unbroken DNA can be isolated from bacteria in a small quantity (Kleinschmidt, Lang and Zahn 1960; Davern, 1966).

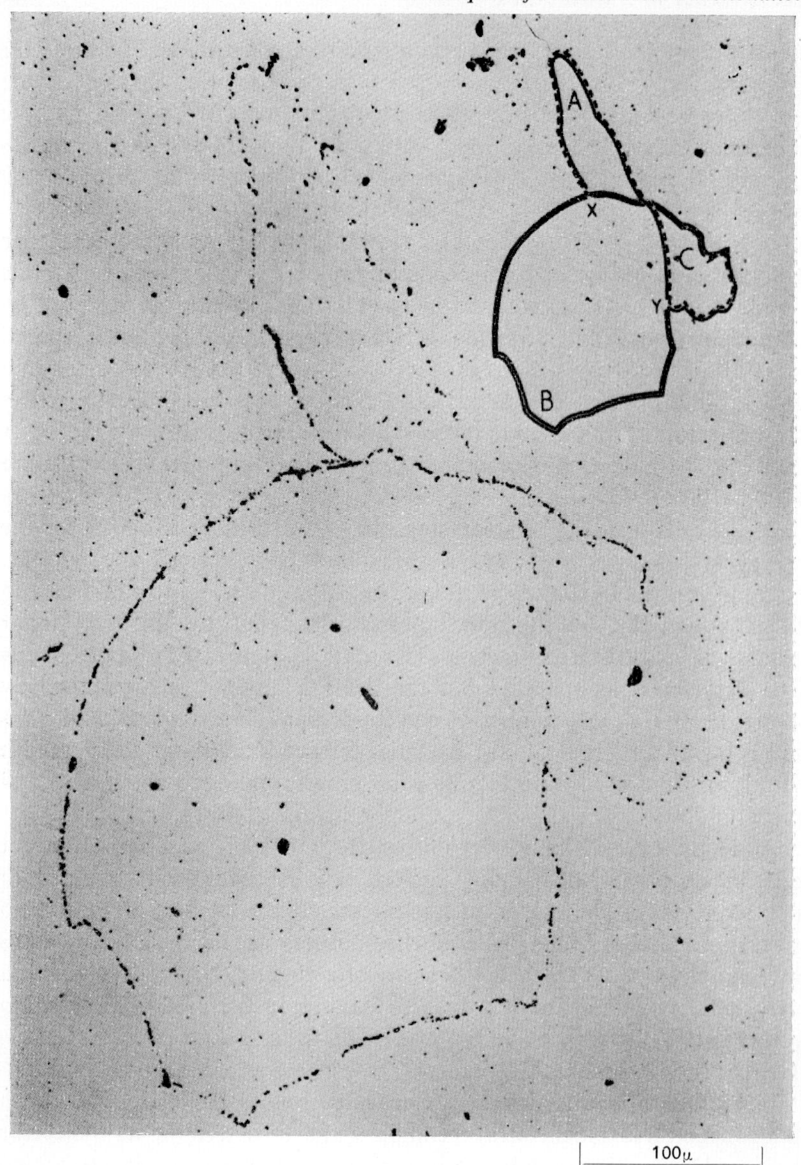

Figure 1.9. Autoradiograph of an *E. coli* K12 Hfr chromosome labelled with tritiated thymidine for two generations and extracted with lysozyme. Exposure time, two months and the scale shows 100 μ. The dashed line, unlabelled strand; solid line, ^{3}H-labelled strand; A and B are the two parts of the chromosome already replicated; C, unreplicated part; Y, growing point; X, swivel. (After Cairns (1963))

Ribonucleic acids

All living cells contain three main kinds of RNA, each of which has a definite role to play in the mechanism of protein synthesis.

(*a*) The bulk of the RNA (about 80 per cent) is contained in cytoplasmic particles known as ribosomes. This ribosomal RNA (rRNA) has a high molecular weight (0·5 to 2·0 × 10^6) and is metabolically stable. The ribosomes are the sites of polypeptide chain synthesis.

(*b*) The next most abundant type (about 15 per cent) is the transfer RNA (tRNA). This has a much lower molecular weight (about 25,000), and is responsible for the transfer of amino acids from the metabolic pool to the ribosomes.

(*c*) The remaining 5 per cent (or less) is metabolically labile, and is referred to as 'messenger' RNA (mRNA) because it functions as a messenger between the DNA and the ribosomes, carrying the coding sequence for the protein chains.

Ribosomal RNA

Ribosomal particles are nearly spherical with a diameter of approximately 200 Å as shown by electron microscopy, and have a sedimentation value of 70S (Tissières, Schlessinger and Gros, 1960; McQuillen, 1965). When the Mg^{2+} concentration is lowered from 10^{-2} to 10^{-4} M the 70S particles dissociate reversibly into two equal sub-units 50 and 30S, both of which are essential for protein synthesis. Furthermore, at high concentrations of Mg^{2+}, the 70S particles aggregate in pairs to form 100S particles (Figures 1.10 and 1.11).

$$100S \rightleftharpoons 70S \rightleftharpoons 50S + 30S$$

In the process of protein synthesis, mRNA and tRNA bind to the 30S sub-unit, while the growing polypeptide chain with attached tRNA is bound to the 50S sub-unit. Several distinct protein fractions from

Figure 1.10. Association of ribosomal sub-units

ribosomes are thought to be responsible in some way for these binding abilities displayed by the respective ribosomal sub-units (for further details see Lewin (1970)).

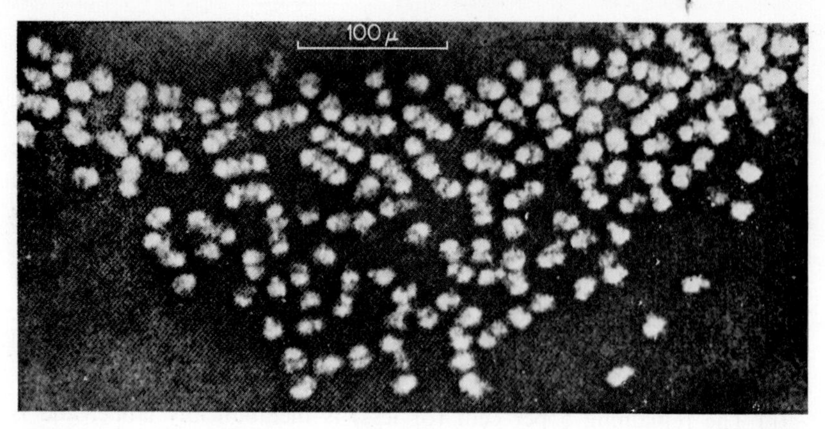

Figure 1.11. Electronmicrograph of *E. coli* ribosomes preparation fixed and 'negatively stained' by drying in a thin film of sodium phosphotungstate. The preparation contains 70S and 100S particles. The 70S particles appear to contain 30S and 50S sub-units. The 100S particles appear to contain two 70S sub-units. (After Huxley and Zubay (1960))

The 50S and 30S sub-units have a molecular weight of $1{\cdot}8 \times 10^6$ and $0{\cdot}85 \times 10^6$, respectively, and both contain 64 per cent RNA and 36 per cent protein (Tissières and coworkers, 1959) in addition to a small amount of polyamines (Spahr, 1962). The dimensions of the particles are:

70S: 20 mμ × 17 mμ
50S: 14 mμ × 17 mμ
30S: 9 mμ × 17 mμ

The 50S sub-unit contains a single RNA molecule of 23S and another rRNA species of about 5S (Galibert and coworkers, 1967), whereas the 30S species contains only RNA of 16S. The molecular weights of the 23S and 16S rRNA's have been estimated using physical (Stanley and Bock, 1965) and chemical methods (Midgley, 1965) and are approximately $1{\cdot}1 \times 10^6$ and $0{\cdot}55 \times 10^6$, respectively. It was originally thought that the 50S sub-unit contained two molecules of 16S rRNA instead of the one 23S molecule (Kurland, 1960), but careful preparations of the 50S sub-unit by Iwabuchi and coworkers (1965) yielded only 23S rRNA and no 16S rRNA.

The base compositions of the two rRNA's, as determined by Spahr and Tissières (1959) show an asymmetrical base composition of high guanine and low cytosine content, but there are significant differences between the two types (Midgley, 1962). They also differ in nucleotide distribution (Aronson, 1962; Takanami, 1967) and the 5′-terminal nucleotide sequences of the 23S and 16S rRNA in *E. coli* are entirely different from each other. This data, together with the evidence of Yankofsky and Spiegelman (1963) that there are two distinct cistrons for the two types of rRNA, would clearly indicate that the 23S and 16S rRNA are independent molecular species.

The 5S rRNA molecule present in ribosomal preparations from both bacterial and mammalian cells was shown to be a distinct ribosomal species (Comb and Sarker, 1967; Brownlee, Sanger and Barrell, 1967, 1968; Forget and Weissman, 1969). The nucleotide sequences of the 5S RNA of *E. coli* and human tumour cells were determined and shown to resemble tRNA in their base composition but contain no methyl bases or pseudouridine. Knight and Darnell (1967) reported that one 5S RNA molecule is bound to the 50S ribosomal sub-unit, and this attachment is permanent (Kaempfer and Meselson, 1968) indicating that it could be a structural component. However, Morell and Marmur (1968) found that treatment with NH_4Cl or CsCl released the 5S rRNA. Brownlee, Sanger and Barrell (1967, 1968) also showed that in *E. coli* there were two 5S rRNA's indicating that more than one gene is responsible for its synthesis.

The chromosomal location of DNA base sequences complementary to 4S tRNA, 5S, 16S and 23S rRNA in *Bacillus subtilis* has been mapped by Smith, Dubnau, Morell and Marmur (1968) using RNA–DNA hybridization (see Chapter 2). It was found that there were 3–4 genes for the 5S RNA, 9–10 genes for the 16 and 23S rRNA and approximately 40 for tRNA. Using a modified density transfer method with germinating *B. subtilis* spores, it was concluded that about 60–80 per cent of the 16 and 23S rRNA and tRNA cistrons is located to the left of the purine B6 marker on the *B. subtilis* chromosome, the remaining 20–40 per cent of these genes being in the terminal portion of the chromosome. The average map position was found to occur in the order: streptomycin resistance, 23S rRNA, 16S rRNA, tRNA and purine B6 whereas the 5S RNA genes were for the most part close to those of the 16 and 23S rRNA loci.

Transfer RNA

Transfer RNA's (tRNA's) which are aminoacylated by the aid of aminoacyl-RNA synthetases, carry amino acids to the proper position on templates. In each cell, there are many tRNA's, each specific for an individual amino acid. A 'codon' (a triplet of nucleotides) in the template selects the appropriate tRNA, not because of the amino acid itself, but because of the nucleotide sequence of the 'coding site' in the tRNA molecule. The idea that an individual tRNA molecule can adapt to a specific code in mRNA, was originally advanced in the 'adaptor' theory by Crick (1957), and was proved experimentally (Chapeville and coworkers, 1962; Weisblum and coworkers, 1962).

There are multiple tRNA molecules even for a single amino acid (Brown, 1963), and each functions preferentially with a specific coding on the template (Bennett, Goldstein and Lipmann, 1963). Most of the coding triplets of mRNA have been established by Nirenberg, Khorana, Ochoa and coworkers (Ochoa, 1963; Nirenberg and coworkers, 1965) but the mechanism whereby the tRNA molecule recognizes its specific amino acid is not completely understood.

Transfer RNA from all species has a molecular weight of about 25,000; a high content of GC and a relatively large proportion of unusual bases.

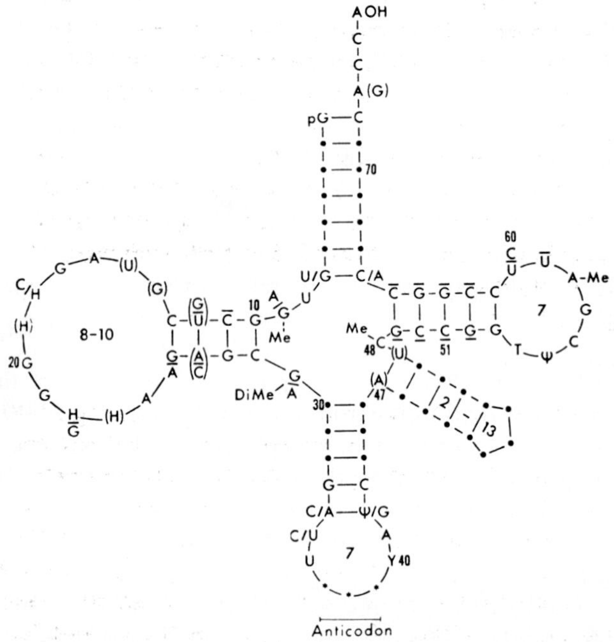

Figure 1.12. Generalized structure of tRNA. The 'dihydro-U' loop is at the left, the 'pentanucleotide' loop at the right. The discrepancy in nucleotides in the dihydro-U loop reflects the fact that in none of the known structures do all nucleotides appear together. Nucleotides designated by a capital letter appear always in this position. Those marked by a bar and a capital letter appear most often, although they might be exchanged for any other nucleotide. Nucleotides in parentheses can, but most do not, occur in this position. Nucleotides indicated by dots are variant. Methyl groups have been included in three positions, where they are found most often. (After Philipps (1969))

Hall (1965) has described eighteen minor nucleosides isolated from yeast tRNA ranging from 2′-*O*-methylpseudouridine (0·0009 mole per cent) to pseudouridine with 4·5 mole per cent.

The nucleotide sequences of about fourteen different tRNA's are known (for further details see Lewin (1970)), and it is generally accepted that the molecules are folded into a definite tertiary structure, which depends on the interactions as observed in protein conformations and nucleic acid double helices (Arnott and coworkers, 1967). The models for this tertiary structure are numerous and all stem from the original clover leaf model of Holley and coworkers (1965). In this model, there are four 'arms' arranged as in a clover leaf (Philipps, 1969) (Figure 1.12). On one arm is the amino acid-acceptor sequence of pCpCpAOH, and opposite it is the anticodon arm with the triplet base-paired regions (Clark, Dube and Marcker, 1968) (Figure 1.13). The anticodon is assumed to have the same double-helical conformation observed for viral RNA (Arnott and coworkers, 1967), and it is the relative orientation of these helical regions which has led to the wide diversity among the models. Recently Levitt (1969) postulated a further model which was very compact, having only eight pyrimidines unstacked. There were 81 hydrogen bonds, and two double-helical regions each of which had a continuous ribose–phosphate chain in one of the strands (Figure 1.14). This model appears to be the best proposed so far, and is in good agreement with many of the experimental observations.

Messenger RNA

Since DNA is physically separate from the ribosomes (the site of protein synthesis), protein cannot be synthesized directly on the genes. An intermediate carrier of information is therefore necessary. Thus Jacob and Monod (1961) postulated the existence of an unstable template in the form of a RNA molecule known as messenger RNA (mRNA). It was originally thought necessary for mRNA to possess certain properties not possessed by other RNA species. First, it must be rapidly synthesized and have a high turnover rate (that is it must be highly unstable once the genetic message has been translated). Secondly, it must have a base composition similar to one strand and complementary to the other strand of DNA.

The existence of a mRNA molecule has been shown in many ways using either bacterial cells, or cells infected with bacteriophages where newly synthesized phage proteins are initiated by the phage DNA. Volkin and Astrachan (1957) transferred phage T_2 infected *E. coli* to a ^{32}P-containing medium, and hydrolysed the ^{32}P-labelled RNA to determine the distribution of radioactivity among the four nucleotides. The ratio A + U/G + C in the RNA (1·7) was similar to the A + T/G + C ratio (1·8) of the phage DNA, but was different from that of the bacterial DNA (1·0), and also from the bulk of the bacterial RNA (0·85). Moreover,

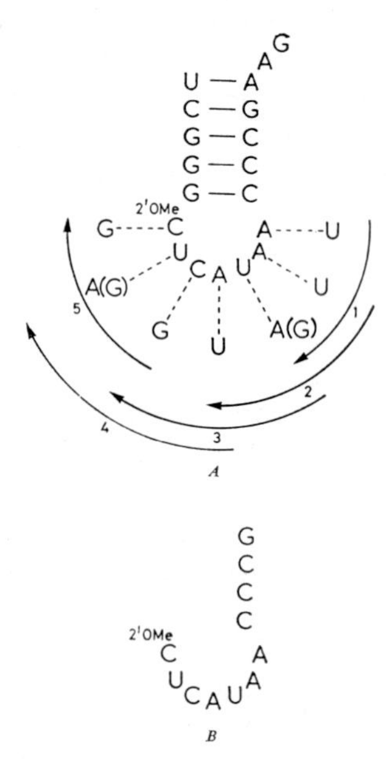

Figure 1.13. The nucleotide sequence of the anticodon loop fragment is shown with letters representing nucleotides in order from the 5′-phosphate towards the diol end. The numbered arrows in *A* indicate possible triplet sequences anti-parallel and complementary with the unpaired region. Complete digestion of the anticodon loop with T_2 ribonuclease releases the single-stranded undecanu-cleotide as shown in *B*. (After Clark, Dube and Marcker (1968))

numerous experiments indicated the ability of mRNA to hybridize effectively with homologous denatured DNA in a highly specific manner, but not with heterologous DNA (Hayashi, Hayashi and Spiegelman, 1963; Mangiarotti and Schlessinger, 1967; Pigott and Midgley, 1968).

The presence of RNA molecule with a short half-life was shown by Levinthal, Keynan and Higa (1962). However, the half-lives of mRNA's vary considerably from species to species, as well as within the same species, so that the original concept that a mRNA should be short-lived is not always true (McCarthy and Hoyer, 1964; Arnstein, Cox and Hunt, 1964).

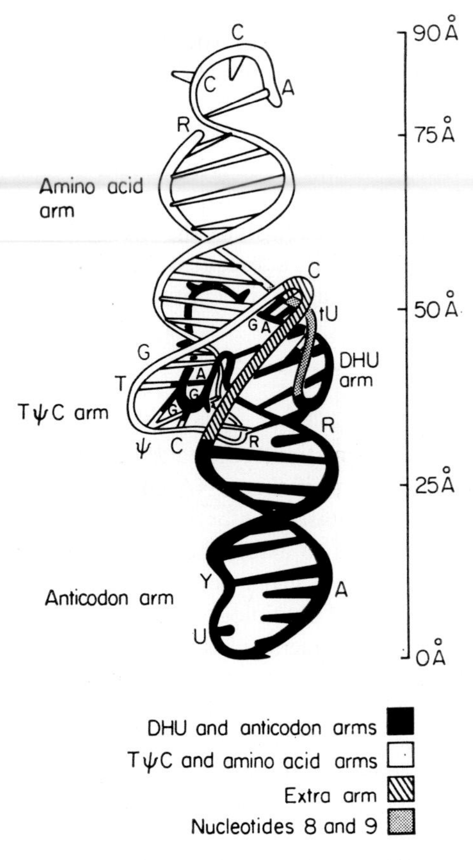

Figure 1.14. A diagram of the tRNA tertiary structure. The letters show some of the common nucleotides. (After Levitt (1969))

B. Biological assays of nucleic acids

Introduction

Each technique described in subsequent chapters fractionates nucleic acids by an essentially different mechanism. Knowledge of both the mechanism of fractionation and nucleic acid structure can be used theoretically to predict the order of elution of the various types of nucleic acids. However, for complete proof of fractionation, each component isolated must be characterized by either reference to known species or by specific assays.

Two of the more extensively used specific biological assays are: (*a*) bacterial transformation for DNA and (*b*) the amino acid-acceptor assay for tRNA.

Bacterial transformation can be used as a tool to study the nature of DNA; that is whether or not the DNA is in the native, denatured or nicked state. It can also be used to study a particular segment of DNA enriched with a particular gene, as well as to study the effect of chemicals and drugs on the structure of DNA and its subsequent function.

The amino acid-acceptor assay indentifies the tRNA molecules for each amino acid.

Bacterial transformation

During bacterial transformation, cells take up exogenous DNA from the surrounding medium, and if this DNA contains genetic information this may under certain circumstances be expressed in the recipient cells. Transformation may be used to assess the biological integrity of a piece of DNA. Not all bacteria in a given population can take up exogenous DNA, only those in the correct physiological state. When their physiological state allows uptake and subsequent transformation by DNA the cells are said to be competent.

Transformation has been observed in a large number of bacterial genera and species (Avery, MacLeod and McCarty, 1944; Spizizen, 1958) (Figure 1.15) and intergeneric transformation has also been reported (Pakula and coworkers, 1962). Inter-specific transformation has been obtained between *Bacillus* species (Marmur, Seaman and Levine, 1963) and *Rhizobium* species (Balassa, 1955) but is always less efficient than intra-specific transfer. Only when the GC ratios of their respective DNA's differ only by 4 per cent or less, is inter-specific transformation possible (Goldberg, Gwinn and Thorne, 1966). The transformation process is not restricted to any particular genes but the characters so far studied are those which can be readily selected.

(a) Uptake of DNA by Competent Cells.

General considerations. Heterologous DNA will readily penetrate competent bacteria, but subsequent incorporation into the host genome is very poor. Schaeffer (1958) suggested that synapsis was prevented because the difference in base sequence between the donor and the recipient DNA's was relatively large. Homologous DNA, on the other hand, has many regions of similarity and the frequency of recombination is correspondingly greater. Ravin (1959) has demonstrated that reciprocal transformation, i.e. mutant DNA transforming wild-type cells to cells with a genetic loss of a particular character, is also possible.

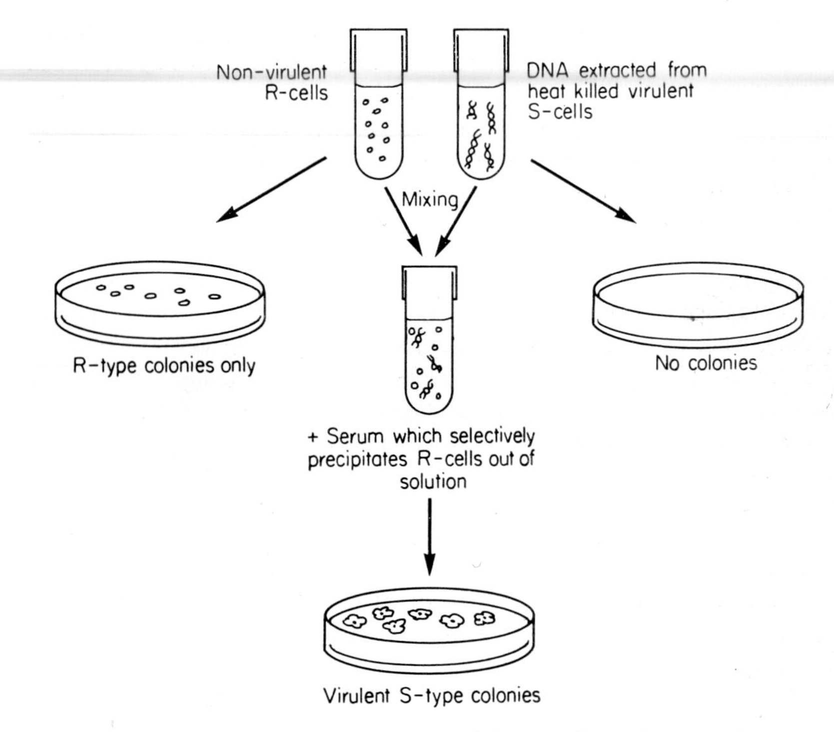

Figure 1.15. Schematic representation of the transformation experiment of Avery, MacLeod and McCarty (1944). DNA extracted from virulent smooth pneumococcal cells (S), is able to transform non-virulent rough pneumococcal cells (R) into virulent smooth type

Mechanism of uptake. DNA uptake can be divided into two main stages; a reversible binding stage which is initiated almost instantaneously (Levine and Strauss, 1965) followed by irreversible binding during which the DNA penetrates the cell (Hotchkiss, 1957). Barnhart and Herriot (1963) showed that the reversible binding was independent of temperature and did not require metabolic energy, since it was not inhibited by dinitrophenol. However, alteration of the DNA structure, either by sonication or denaturation, considerably reduced the reversible binding. Significant alterations in binding could also be observed if the ionic strength or pH of the medium was altered. In general, increasing the ionic strength ten-fold considerably reduced binding, while lowering the pH to 5·5 increased binding, even to denatured DNA (Postel and Goodgal, 1966). This suggests that the initial DNA binding may be electrostatic in nature.

Irreversible uptake is measured by resistance of the DNA to deoxyribonuclease, and is strongly dependent on temperature (Ayad, 1964). It requires metabolic energy (Stuy and Stern, 1964) and no protein synthesis seems to be required (Young and Spizizen, 1963). It has been suggested by Levine and Strauss (1965), that DNA molecules are taken up in such a way that the DNA enters a deoxyribonuclease-insensitive zone in a lengthwise manner.

Site of DNA uptake. This has been investigated by several workers. The cytoplasmic membrane beneath the cell wall was implicated by Young (1967), who, using autoradiography, observed that labelled transforming DNA was visibly absorbed near the wall–membrane complex. Wolstenholme, Vermeulen and Venema (1966), also using autoradiography, found that absorbed DNA was closely associated with certain invaginations of the membrane, the mesosome organelles. Moreover, the number of mesosomes in competent cells was observed to increase to approximately two to three times the number found in cells taken from earlier stages of growth.

The idea that mesosomes are involved in DNA uptake is supported by Tichy and Landman (1969) who found that transforming DNA could become associated with the surface of spheroplasts, and that uptake was only achieved after cell-wall formation had been resumed. Since (*a*) mesosomes could be shown to have been everted from the spheroplasts, and (*b*) cell-wall formation may be associated with mesosome activity, the incorporation of DNA at this stage could possibly be connected with the reformation of mesosomal vesicles. If cells were grown on a 25 per cent gelatin medium, 75 per cent of the cells lacked mesosomes, because of their expulsion into the space between the wall and the membrane, and this resulted in a very low yield of transformants.

Autoradiographic studies on DNA uptake by *B. subtilis* carried out by Javor and Tomasz (1968), provided evidence that the irreversibly absorbed DNA remained at certain fixed sites within the cell. These sites, which were occupied for at least 90 min., were localized at either the poles or midpoints of the bacterium. Since it is known that these positions are associated with the process of cell division, it was suggested that the organelles, normally involved in cell division, might be participating in DNA uptake. Ryter (1967) has examined many electronmicrographs of dividing *B. subtilis* and concluded that the mesosome is bisected during chromosome separation, each new mesosome transporting a daughter chromosome on division (Figure 1.16). It has been proposed by Akrigg,

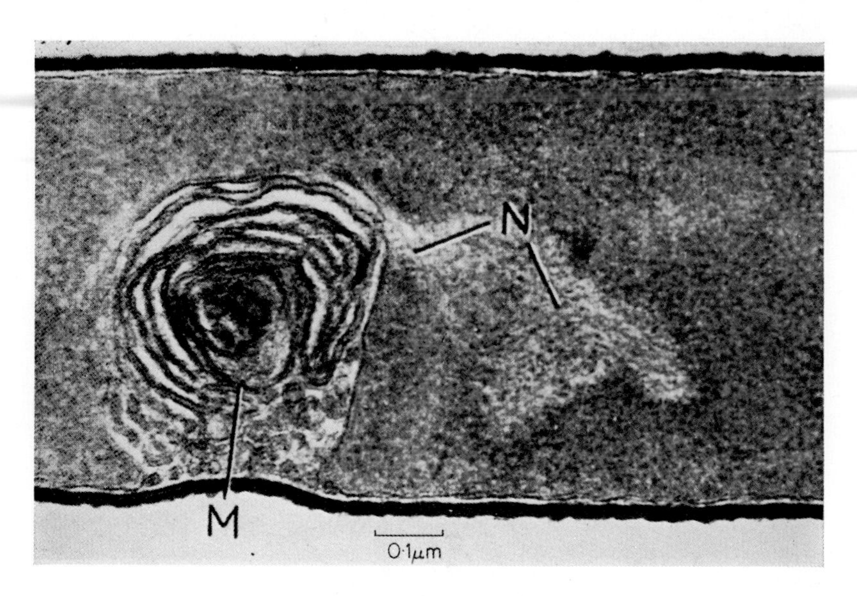

Figure 1.16. Electronmicrograph of ultrathin section of *B. subtilis*, showing the attachment of the nucleus (N) to the mesosome (M). (After Ryter and Jacob (1964))

Ayad and Blamire (1969) that uptake of DNA by competent cells of *B. subtilis* occurs by way of the mesosome, at a time when the cell wall of the bacterium has been weakened due to the action of a lytic enzyme (Akrigg and Ayad, 1969, 1970) just prior to, or just after, cell division (Figure 1.17).

(b) **Integration of Donor DNA.** Once the donor DNA has penetrated the bacterium, it must interact with the host chromosome or any replicon so that the donor markers can be genetically expressed in the same manner as the host markers. Two mechanisms have been postulated to explain how integration and recombination occur and are usually referred to as 'copy choice' and 'breakage and reunion' (Lederberg, 1955). Either the donor DNA acts as a template in the replication of the host DNA, but

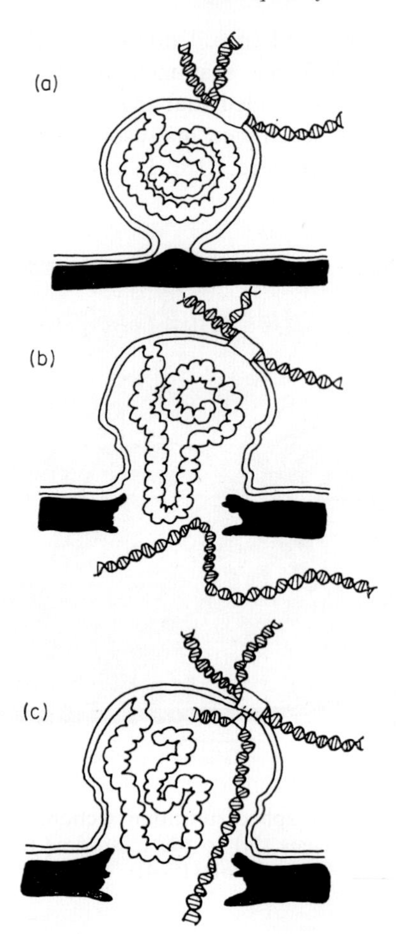

Figure 1.17. (a) Intact mesosome just before competence. The site of attachment of the DNA to the membrane is now the replicating point
(b) The lytic competence factor has acted on the cell wall leaving a 'hole' through which the mesosomal vesicles can protrude. This forms an'active site' for the attachment of exogenous DNA.
(c) The transforming DNA is now inside the mesosomal invagination and has come in contact with the replicating point where integration can occur. (After Akrigg, Ayad and Blamire (1969))

is not physically inserted, or part of the donor DNA molecule is inserted into the host chromosome by a process involving fission and repair, as postulated by Bodmer and Ganesan (1964) and shown in Figure 1.18.

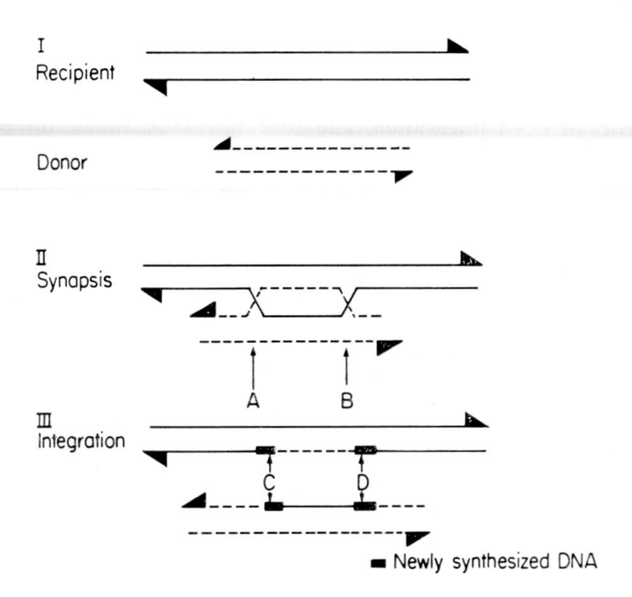

Figure 1.18. Schematic model for synapsis and integration during transformation. (I) Intact donor and recipient molecules. (II) Synapsis by segmental interchange of hydrogen bonding between donor and recipient strands. (III) Integration by breakage of strands (A, B) followed by repair of unpaired regions (C, D) using the unbroken strand as template and covalent linkage to recipient (donor) material. (After Bodmer and Ganesan (1964))

The experimental evidence tends to favour the breakage and reunion theory, as the introduction of donor markers into the recipient genome could be detected in the absence of any measureable DNA replication (Fox, 1960; Voll and Goodgal, 1961). This agrees with the work of Bodmer (1965) and of Archer and Landman (1969a; 1969b) who, using amino acid starvation, were able to develop competent cells of *B. subtilis* in the absence of cell division. The chromosome of these cells was found to be uniformly aligned at the terminus, but this did not prevent the integration of markers whose corresponding recipient locus was on a part of the chromosome distant from the replicating point. Erickson and Braun (1968) pointed out, however, that these results did not exclude the possibility that membrane-bound deoxyribonuclease-insensitive DNA may become integrated at the replicating point, when DNA replication was resumed. The replicating point, as a possible site for integration was first suggested by Bodmer (1965) whose experiments indicated that donor DNA was integrated at the site at which host DNA synthesis occurred.

(c) **Kinetics of DNA Integration.** The rate of DNA taken up by competent *B. subtilis* cells has been investigated by Ayad and Barker (1969), who used 5-bromo-^{3}H-uracil labelled DNA to transform auxotrophic strains of *B. subtilis*. When the DNA was reextracted from the transformed culture after various times of incubation, and subjected to equilibrium sedimentation in caesium chloride gradients, it was found that, in addition to heavy donor DNA and light recipient DNA, there was present a component with an intermediate density which when isolated was found to possess transforming activity. Denaturation of the component gave only one peak on recentrifugation, indicating that integration had occurred in both strands of the recipient DNA. Furthermore, the integration process was found to be inhibited by treating the donor DNA with acriflavin (Ayad, 1969), naladixic acid and ethidium bromide (Ayad, 1970). (For further details on this subject see Hayes (1968).)

Transformation procedure. There are a number of methods for preparing competent cells from different strains of bacteria and the one described here is mainly for *B. subtilis* 168 I^{-} (indole- or tryptophan-requiring mutant).

Growth media: The growth media were based on the glucose minimal medium described by Spizizen (1958) containing the following per litre: K_2HPO_4, 14g.; KH_2PO_4, 6g.; $(NH_4)_2SO_4$, 2 g.; trisodium citrate. $2H_2O$, 1g.; $MgSO_4.7H_2O$, 0·2g. and glucose (sterilized separately) 0·5g.

The minimal medium was supplemented with $MnSO_4$ (10^{-7} M) and the following 20 amino acids (20 μg./ml. each) added aseptically: glycine and L-amino acids alanine, aspartic acid, arginine, asparagine, cysteine, cystine, glutamic acid, histidine, isoleucine, leucine, lysine, methionine, phenylalanine, proline, serine, threonine, tyrosine, tryptophan and valine. L-Tryptophan was prepared, sterilized and added separately. This medium will be called medium I.

Sterilization: Minimal medium containing agar, and glucose were sterilized separately as described above. Stock solutions of amino acids were sterilized by filtration (0·45 μ, 27mm. Oxoid membrane filters).

Pyrex petri dishes, test tubes and pipettes plugged with cotton wool, were placed in metal canisters or cardboard tubes and sterilized by heating in an oven at 160°C for 2h.

Preparation of agar plates: Pyrex petri dishes (94mm.) or pre-sterilized plastic petri dishes (90mm) were used.

(*a*) Transformation plates: The agar medium used to assay the number of transformants was based on Spizizen's glucose minimal medium (Spizizen, 1958) with 2 per cent (w/v) Difco Bacto-agar, supplemented by the following amino acids (50 μg./ml. each), glycine and L-amino acids of aspartic acid, cystine, isoleucine, leucine, lysine, methionine, proline, serine and threonine (Levine and Strauss, 1965).

(*b*) Viable count plates: To assay the number of viable cells in a culture, the supplemented agar medium described above was further supplemented with L-tryptophan (20 μg./ml.).

Preparation of Competent Cells: This was carried out using the method described by Akrigg, Ayad and Barker (1967) and Akrigg and Ayad (1970). Cells of *B.subtilis*, 168 I^- were inoculated from a slope into medium I to give an extinction of 0·3 at 600 mμ using a Unicam, SP 600 spectrophotometer. This culture was incubated at 37°C for 4h. with vigorous aeration, and was subsequently diluted into 1l. of medium II. Medium II differed from medium I in containing $MnSO_4$, L-tryptophan at the reduced concentration of 5 μg./ml., and in addition α,α′-dipyridyl (Sigma Chemicals) (40 μg./ml.). α,α′-Dipyridyl probably stimulates competence by forming complexes with cupric ions, which have been found to inhibit competence (Anagnostopoulos and Spizizen, 1961). The cells were incubated in medium II at 37°C for 60 min. with slow aeration.

Transformation assay: The resulting competent culture was tested for frequency of transformation as follows: 5ml. samples of the culture in test tubes (12·5 × 1·9 cm.) were incubated with DNA (5 μg./ml.) at 34°C for 90 min. with vigorous shaking (reciprocal shaker), followed by incubation at 37°C for 5 min. with deoxyribonuclease (Sigma Chemicals) at a concentration of 50 μg./ml. to destroy excess DNA. The cell suspension (1·0 ml.) was then serially diluted with 10 ml. aliquots of glucose minimal medium, and the transformants scored by spreading 0·2 ml. samples of each dilution on transformation plates, followed by incubation at 37°C for 24h. The number of viable cells on the culture was scored by plating 0·2 ml. samples of the same dilution on viable count plates, followed by incubation at 37°C for 24h. All platings were duplicated. Controls for contamination, and reversion to prototrophy of the mutant, were carried out with each experiment (Figure 1.19).

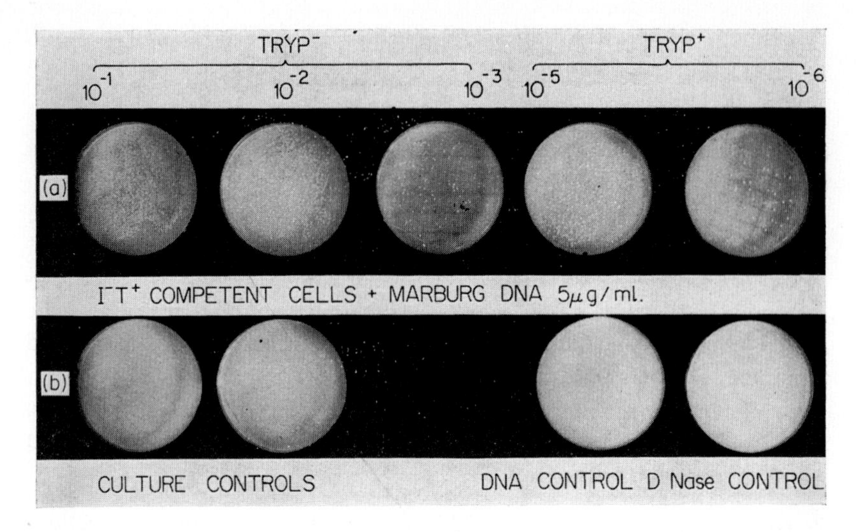

Figure 1.19. Photograph illustrating transformation of *B. subtilis*. Competent cells of *B. subtilis* 168I⁻ (indole requiring mutant) were incubated with DNA, isolated from the Marburg strain of *B. subtilis* at a final concentration of 5 μg./ml. and incubated for 30 min., followed by incubation with deoxyribonuclease and Mg^{2+} (see text). (a) Culture plated, after various dilutions, on tryptophan-lacking plates to score for the transformants and on tryptophan-containing plates to score for the total number of viable cells. (b) Controls used for the transformation assay (culture, DNA and DNase)

The amino acid-acceptor assay

This method of characterizing a particular tRNA involves the incubation of one radioactive amino acid (whose specific tRNA is being assayed) and nineteeen non-radioactive amino acids with a tRNA mixture in the presence of all the factors reuqired to link the amino acids to their respective tRNA molecules. The amount of radioactivity incorporated into RNA is then a measure of the amount of amino acid-specific tRNA originally present in the mixture. In this way, a solution can be assayed, for example, for valine-accepting activity by using radioactive valine and other non-radioactive amino acids.

(a) **Isolation of the activation enzyme.** This enzyme attaches the amino acids to the tRNA molecules, and is prepared from either growing cultures of yeast cells, or from frozen cells which have been stored in liquid nitrogen, by the method of Cherayil and Bock (1965). The cells which possess a strong cell wall are difficult to break open. Cherayil and Bock (1965) mixed the cells (50 g.) with a suspension of glass beads (diameter 120 μ) in 30 ml. of tris-HCl buffer (0·01 M) pH 7·5 containing 0·01 M magnesium acetate, 0·03 M ammonium chloride and 0·005 M β-mercapto-ethanol. This mixture was ground for 15 to 20 min. to break open the cells, and the cell debris and glass beads removed by centrifugation at 2,000 g. for 30 min. Other cellular particles, such as ribosomes, were subsequently removed by a further ultracentrifugation of the clear supernatant at 40,000 r.p.m. for 3 h., in a Spinco Model L ultracentrifuge. The supernatant was then dialysed for 16–20 h. against the buffer used to grind the cells. All operations were carried out at a low temperature in order to decrease any proteolytic digestion, and to ensure a highly active enzyme preparation. Following dialysis, the non-diffusible material was quickly frozen in ethanol-dry ice and stored at a low temperature using nitrogen.

(b) **Assay Procedure.** The following assay of amino acid-acceptor activity is the micromodification of a more usual assay procedure, and has the advantage that only very small quantities of material are required, and large numbers of samples can be simultaneously analysed. An aliquot (0·05 to 0·15 ml.) of each solution to be treated (usually fractions from a chromatographic separation of tRNA's), is applied directly to small circles of filter paper (Whatman 3 MM, 2·3 cm. in diameter) which have previously been marked for later identification. The discs must not of course be in contact with each other, and for accuracy it is necessary for them to have the minimum contact with the bench or support. This is achieved in several ways, one of which is to have a block covered with a nylon or teflon sheet, into which rows of pins have been placed. The discs can then be supported by the pins with the minimum external contact. The discs containing the samples are dried by passing a slow current of air over them, and then transferred to cold 75 per cent ethanol containing 0·03 M KCl. This precipitates the tRNA molecules on the discs and they can be washed free of contaminating substances (salt, etc.) with the cold ethanol–KCl mixture. The discs are dried and the RNA on each disc can then be incubated, usually at room temperature, with a mixture consisting of 40 μmoles Tris buffer, pH 7·5, 4 μmoles $MgCl_2$, 0·5 μmole EDTA, 1 μmole ATP, 5 μmoles mercaptoethanol, 40 μmoles KCl, 0·3 μc ^{14}C-amino acid of specific activity 10 to 25 μc/mole, 0·01 μmole each of the 19 other amino acids (non-radioactive) and 0·1 ml. of the activating enzyme solution. 0·1 ml. of the total mixture (1·0 ml.) is then added to each disc. In order to prevent the evaporation, and hence premature drying out of each disc, the block containing the samples is covered with a glass dish in which is placed a moist tissue or similar humidifier. After incubation (20 to 30 min.), the discs are dried in a current of air and the sample precipitated by placing the discs in 10 per cent trichloroacetic acid (TCA). As a precaution this, and all other washing solutions, should contain the amino acid (^{12}C) under investigation. The discs are successively washed in 66 per cent cold ethanol containing 0·5 M NaCl, 10 per cent TCA, 5 per cent TCA and finally in an ethanol–ether (3 : 1 v/v) mixture. They are then dried under a lamp, and the radioactivity on each disc determined by placing in a vial containing liquid scintillator and counting in a liquid scintillation counter. The maximum amount of tRNA that could be assayed by this method was 50 μg. per paper disc.

CHAPTER 2

Analytical methods in nucleic acid biochemistry

Introduction

In Chapter 1 the nucleic acids were discussed from the point of view of their structure. This chapter will describe the techniques used to elucidate their physical properties. Many of the more valuable methods need elaborate equipment, and the theory behind the operation of certain machines often requires a level of mathematical knowledge greater than that possessed by most biochemists. For this reason, only a general outline of each technique will be described, but an attempt will be made to show how each may be applied, and the kind of results to be expected.

Microscopy

Direct observation provides information concerning the size and shape of an object, in addition to giving a guide to its structure and function. The observation of biological materials is aided by microscopes, the resolving power of which depends on the wavelength (λ) of the radiation used, the maximum resolution distance being $\lambda/2$. Many attempts have been made to improve the resolving power by reducing the wavelength of the incident radiation, but although ideal in theory, many difficulties are encountered in practice.

Microscopes using ultraviolet radiation (of shorter wavelengths than visible light) have been used to study the cellular location of nucleic acids. The resolution is increased and, since nucleic acids (and proteins) have a high u.v. absorption, they show contrast without special staining. U.v. microscopes, however, cannot distinguish individual DNA and RNA molecules, but they can be used to locate nucleic acids within cells (Caspersson, 1941).

An electron beam has a very short wavelength, (of the order of 10^{-9} cm. or 0·1 Å unit), and should theoretically be capable of resolving reasonably large molecules. For this reason, the technology of controlling electron beams in the same way that beams of light can be controlled, has been extensively studied in order that they may be used in a form of microscope. Electron beams cannot be 'bent' by the usual optical devices but, because of their charge, they can be deviated by magnetic fields. In the electron microscope, therefore, the 'lenses' are coils of wire through which an electric current is passed to produce the required field, and the focal length may therefore be changed at will by altering the electric current. There are, however, three main difficulties: first, magnetic lenses can only be made to converge radiation, and aberrations cannot be corrected by diverging lenses; secondly, the incident electron beam is scattered, and not absorbed, by the object. An electron has only to deviate slightly before it fails to reach the photographic plate. The third difficulty is that the scattering power of an object is determined by the electron density within the object itself. The contrast produced in the electron microscope will therefore depend on differences in density between one part of an object and another. Most biological materials have only small variations in this type of density, and therefore contrast must be increased by special means.

In the case of light microscopy, structures within a tissue slice can be distinguished by staining with specific dyes. The electron microscope cannot detect colour differences, but sections can be stained with heavy-metal compounds such as osmium tetroxide, phosphotungstic acid or uranyl acetate. These act in the same way as other stains, except that they provide contrast by altering the electron density in the areas to which the stain has bound (Brenner and Horne, 1959; Kellenberger and co-workers, 1965).

Difficulties also arise from the fact that electron microscope samples must be mounted in a vacuum, to avoid the scattering of the electron beam by air. Unfortunately, the vacuum causes evaporation, so all water must be removed before observation is possible. This can give rise to distortions unless special precautions are taken, and the possibility that an alteration in structure has occurred during the removal of water cannot be completely excluded. However, despite these limitations, electron microscopy has been used widely to examine nucleic acids both *in situ* in tissue slices, and also in an isolated state.

A monolayer technique for the examination of nucleic acids has been developed by Kleinschmidt and coworkers (Kleinschmidt and Zahn, 1959; Lang, Kleinschmidt and Zahn, 1964). DNA or RNA in solution is made into a monolayer by adsorption to the basic groups on a surface-denatured protein monolayer film. Many basic globular proteins have been used, the most usual being cytochrome c. There are three main modifications of this technique, which essentially converts the nucleic acid from a three-dimensional to a two-dimensional state, known as (1) spreading, (2) diffusion and (3) 'one-step' release procedures.

In the spreading procedure, a solution containing both the protein and nucleic acid is floated on to the surface of an aqueous salt solution, and adsorption occurs during the spreading of the monomolecular protein film. In the case of diffusion, the protein monolayer is formed on the surface of the nucleic acid solution, and the nucleic acid diffuses and is irreversibly adsorbed to the lower surface of the protein film. The shearing forces exerted on the nucleic acids are less extensive in the diffusion procedure than in the spreading technique. However, if spreading is carried out slowly, the two modifications produce comparable electron micrographs. In the 'one-step' release procedure, the monomolecular protein film and the extraction of the nucleic acid occur simultaneously. The protein and virus particle or bacterial protoplast in a solution of high salt concentration are floated on to a solution of low salt concentration. The nucleic acid is therefore released from the virus or protoplast by osmotic shock and subsequently adsorbed to the protein monolayer.

The protein–nucleic acid monomolecular layer is transferred by surface tension to a grid covered with a support film (usually carbon) by horizontally touching the monolayer. A strong adsorption (chemisorption) occurs, and the grid is subsequently shadowed or stained. To shadow, a filament of a heavy-metal atom (e.g. platinum, palladium or uranium) is heated in a vacuum until the individual atoms have sufficient energy to break away from the filament and disperse into the container (see Figure 2.1). Since they are in a vacuum, they will travel in straight lines until they encounter some obstacle, on which they will impinge and come to rest. If an object is placed at an inclined angle to the path of such a stream of heavy atoms, then all the accessible regions will gradually become covered with a thin layer of the metal, but those parts of the object distal to an elevated area will be protected, and will not receive a metal layer. When the object is subsequently viewed in an electron beam, the micrograph is dark where no metal was deposited, and light elsewhere. This gives the effect of an object casting shadows and, from the shape of the shadow, deductions can be made about the outline of the object (Figure 2.2).

Using the 'one-step' release procedure, followed by platinum shadowing, the DNA in the bacteriophage T_2 chromosome had the appearance of a single thread with two free ends, of length 49 ± 4 μ (Figure 2.3) (Kleinschmidt and coworkers, 1962). MacHattie, Berns and Thomas (1965) investigated the DNA isolated from *Haemophilus influenzae* by osmotic shock. The DNA was shadowed in all directions with uranium oxide by rotating the grid, and subsequent electron microscopy revealed a linear, double-helical DNA molecule of length 832 μ.

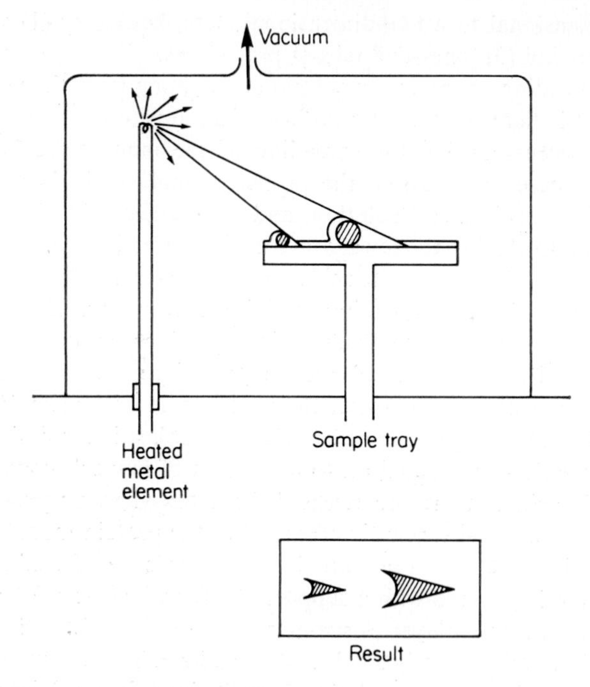

Figure 2.1. Diagrammatic representation of the metal shadowing technique

In addition to determining the length of linear DNA molecules, electron microscopy has demonstrated closed circles in several types of DNA. Most of the replicative form of phage Ø-X174 had a characteristic double-helical appearance, and was circular, having an average contour length of 1·64 μ (Kleinschmidt, Burton and Sinsheimer, 1963). Electron microscopy studies of phage Ø-X174 also enabled a model for the genetic recombination of the circular DNA to be proposed (Rush and Warner, 1968). Cells of *E. coli* infected with the phage were lysed and the DNA prepared, denatured and chromatographed on Sephadex G-100. The excluded fractions were pooled and passed through nitrocellulose, which retained single-stranded DNA (see p. 45). Examination of the eluate by electron microscopy showed that 95 per cent of the DNA was in the form of covalently closed circular duplex molecules, with a contour length twice that of the normal replicative form circular Ø-X174 DNA. During genetic recombination, therefore, single circles were converted to double circles and similar conclusions were made from studies on the replication of bacteriophage λ (Weissbach, Bartle and Salzman, 1968).

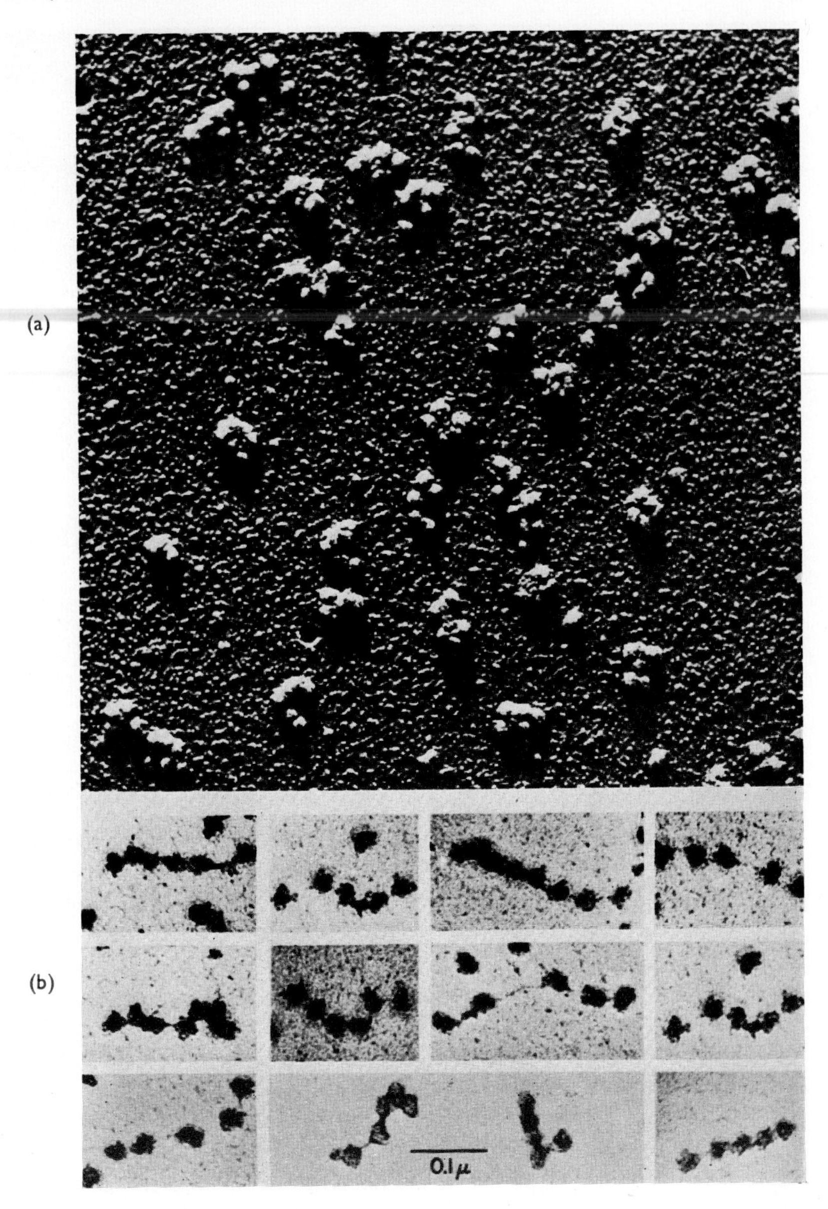

Figure 2.2. Electronmicrograph of reticulocyte polysomes. (a) Platinum shadowed preparation and (b) uranyl acetate staining which shows a faint dark staining thread passing between the ribosomes. (After Slayter and coworkers (1963))

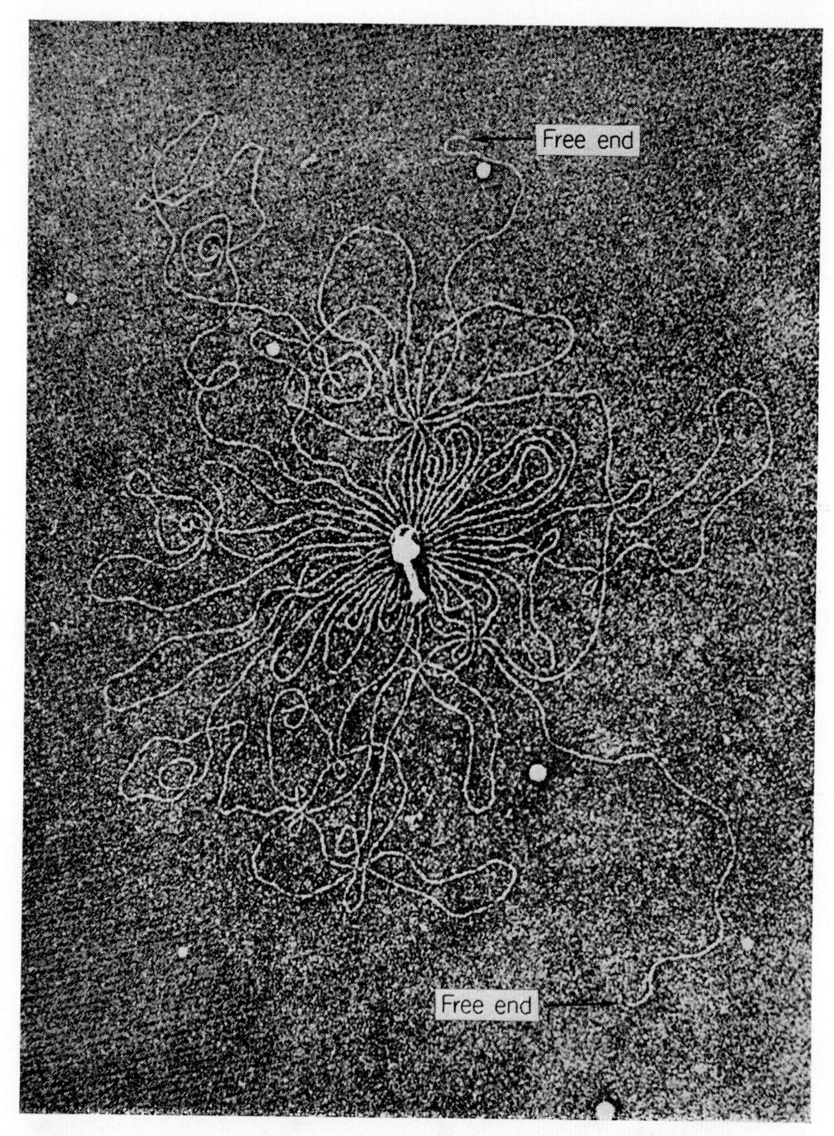

Figure 2.3. Platinum shadowed electronmicrograph of bacteriophage T_2 DNA released by osmotic shock, showing a single thread with two free ends and a length of 49 μ. (After Kleinschmidt and coworkers (1962))

MacHattie and Thomas (1964) and Caro (1965) investigated the unbroken and untangled DNA of phage λ and its deletion mutant λb_2. The molecules of wild-type phage had an average length of 17·3 μ, corresponding to 50,000 base-pairs, and a molecular weight of 33×10^6 daltons, whereas the length of the mutant DNA was 13 μ. These results agreed with those obtained by density-gradient centrifugation, and indicated that the mutant had lost 23 per cent of the wild-type DNA.

Closed-circular DNA exists in mitochondria (Radloff, Bauer and Vinograd, 1967). HeLa cell mitochondrial DNA was centrifuged in a caesium chloride density-gradient in the presence of ethidium bromide (see p. 61) and the fractions examined by electron microscopy. Most of the closed-circular DNA molecules had a mean length of 4·81 μ but there was also a heterogeneous population of smaller DNA molecules of length 0·2 to 0·3 μ.

Electron microscopy showed that 40 to 60 per cent of the mitochondria in mouse fibroblasts (L-cells) was in the form of covalently-linked closed-circular monomers with a contour length of 4·74 μ and a molecular weight of $9{\cdot}1 \times 10^6$ daltons (Nass, 1969). A further ten per cent were multiple-length forms containing two to four interlocked monomers, while the remainder consisted of nicked circles with a few linear fragments. Circular mitochondrial DNA from Ascites tumour, human liver and rat liver cells had comparable contour lengths of 4·75 μ, 5·06 μ and 4·9 μ, respectively.

The closed-circular component of L-cell mitochondria was found to resist denaturation by both heat and alkali. Electron microscopy also showed that it had 33 ± 3 tertiary turns per molecule or 3·6 turns per 10^6 molecular weight. Two bands were observed after CsCl centrifugation, and electron micrographs showed that the fast band consisted of highly twisted monomeric and some dimeric forms, whereas the slower band contained open duplex monomers, single-stranded circles and also linear strands, either free or partially attached to the single-stranded circles.

Supertwisted circular DNA was observed by Bode and MacHattie (1968) after lysogenic infection of *E. coli*. The newly synthesized DNA was purified by sucrose gradient sedimentation and examined by electron microscopy both in low and high salt media. In low salt, a greater percentage of the molecules was superhelical, whereas in high most of the molecules were relatively untwisted (Figure 2.4). It was suggested that the number of base-pairs per 360° turn of the helix (pitch) varied with ionic strength, and that the effect of high ionic strength on the DNA helix was similar to that of the intracellular environment.

Electron microscopy can therefore be extremely useful for examining DNA and distinguishing between its various forms (linear, circular and superhelical). This can be extended to genetic mapping. If one could obtain a hybrid duplex between a wild-type DNA strand and a deleted mutant strand, and examine the duplex using electron microscopy, one might be able to observe the position of gene deletion in the mutant strand. Preferential isolation of the true hybrid could be achieved by hybridizing (see p. 43) heavy labelled wild-type DNA (^{15}N and D_2O) with

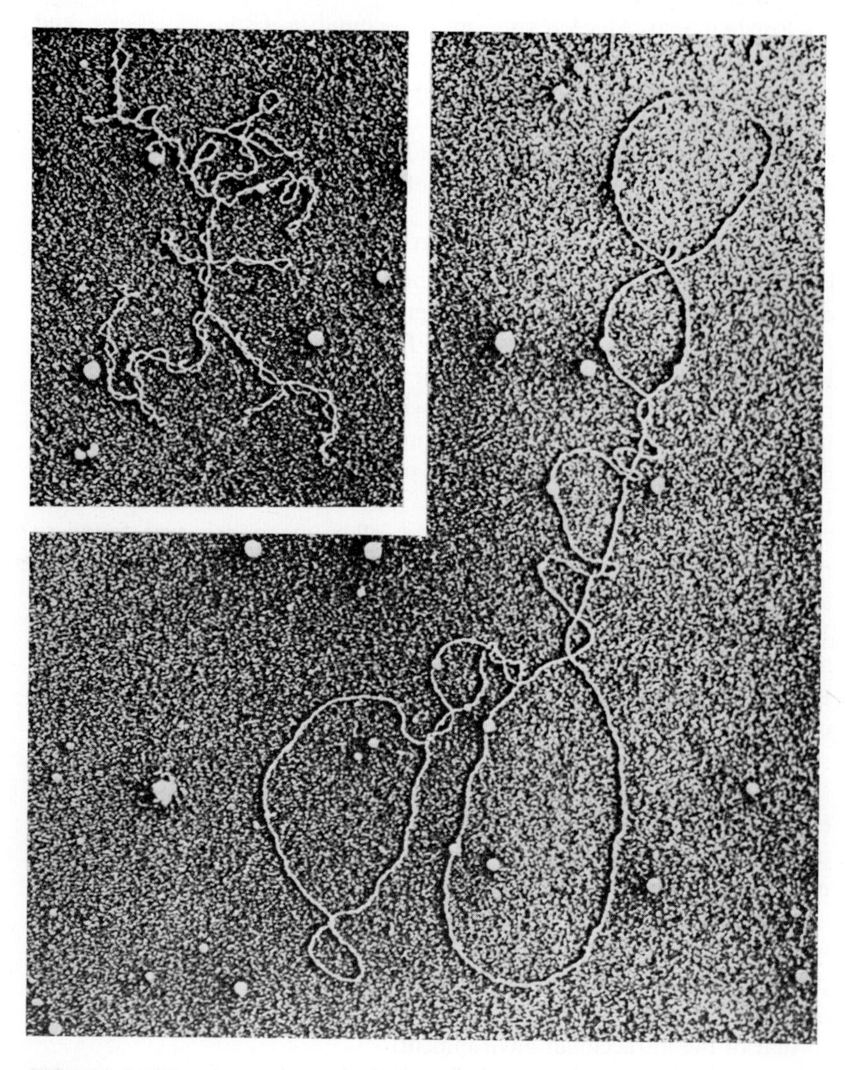

Figure 2.4. Examples of twisted (128 crossings) and relatively untwisted (22 crossings) circular λ DNA molecules. (After Bode and MacHattie (1968))

light mutant DNA (^{14}N and H_2O) or vice versa, followed by caesium chloride density-gradient centrifugation (see p. 56).

Autoradiography

Another way of observing nucleic acid molecules directly is by autoradiography. In this technique, radioactive tritium (^{3}H) (a β^--emitter) is incorporated into the molecules under investigation, usually by growing the organism in the presence of tritiated compounds unique to the particular macromolecule, e.g. thymidine for DNA and uridine for RNA. These precursor units become a normal functioning part of the macromolecule, which is then isolated and placed in a suitable sensitive photographic emulsion. The emulsion is subsequently stored for some time in the dark, during which emissions from the ^{3}H-molecules cause silver to be deposited in the emulsion. It can then be developed in a similar way to ordinary photographic plates, when the silver shows up as black grains. The presence of the labelled compound can be detected by a clustering of these grains in a certain area. This particular method has proved very valuable in tracing the progress of certain macromolecules during particular biological processes. A particularly elegant example of the use of autoradiography is the work of Cairns (1961, 1963), who used the method to show the circularity of bacterial DNA (Figure 1.9) and the structure of T_2 bacteriophage DNA (Figure 2.5).

Autoradiographic techniques have also been used to determine the molecular weights of very large DNA molecules. In this example, a high concentration of ^{32}P is incorporated into T_2 bacteriophage DNA. The whole particle is then embedded as described previously in sensitive emulsion, and the radioactivity allowed to decay for a suitable length of time. When the plate is developed, the number of β-ray tracks emanating from a single point are measured. These points correspond to the bacteriophage DNA. Each track represents the decay of a single ^{32}P atom, and the total phosphorus content of each bacteriophage DNA molecule can be estimated from the number of tracks, the known half-life of ^{32}P, the exposure time and the experimentally determined specific activity of the bacteriophages used. This method has given values of $1{\cdot}3 \times 10^8$ daltons for bacteriophage T_2 DNA (Davison and coworkers, 1961; Rubenstein, Thomas, Jr. and Hershey, 1961).

Hypochromism

The common heterocyclic bases absorb u.v. light of a certain wavelength, and typical absorption spectra for the five most common bases are shown in Figure 2.6. The intensity of this absorption is useful not only

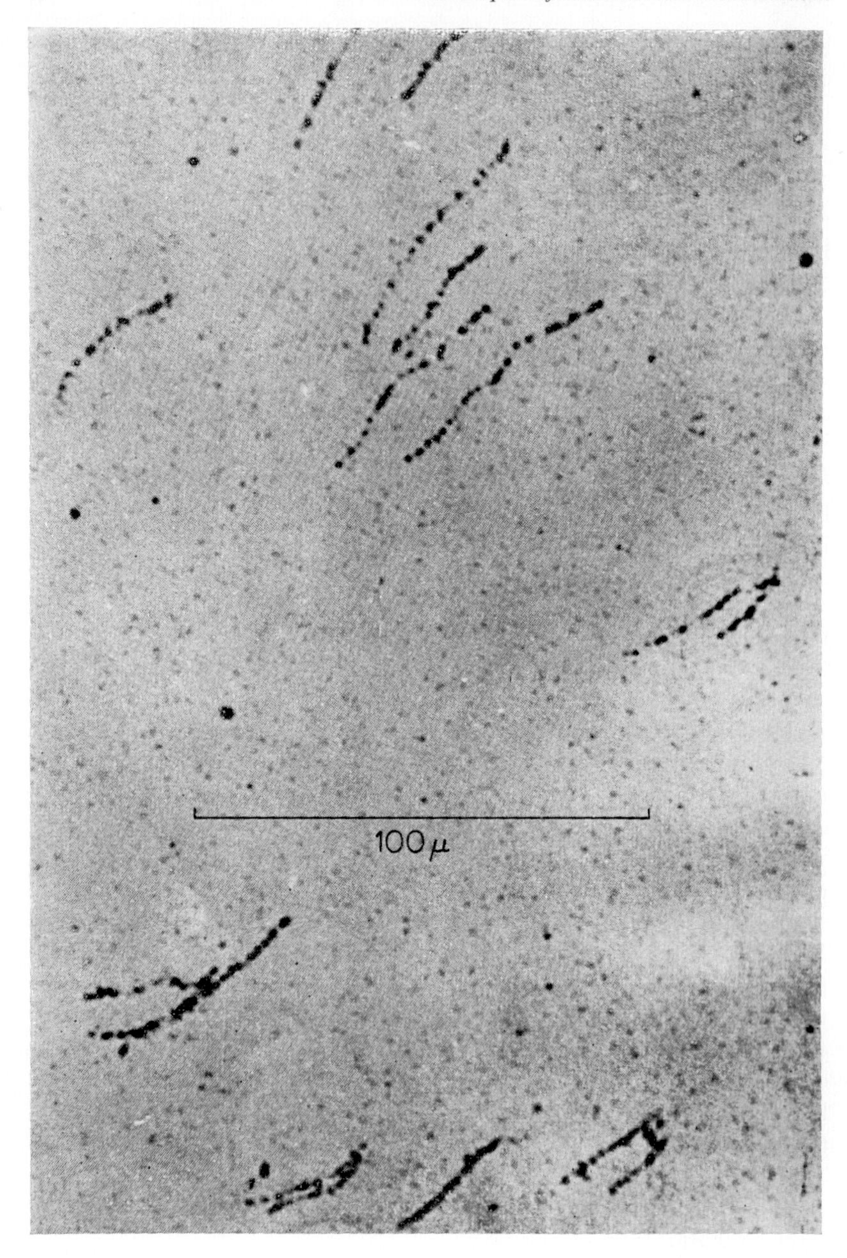

Figure 2.5. Autoradiograph of bacteriophage T_2 DNA which has incorporated ^{3}H-thymidine. The scale is 100 μ long. (After Cairns (1961))

for identifying the bases, but also as a rapid way of ascertaining the presence of nucleic acids in chromatograms or electrophoretograms. These can be 'developed' by irradiating the paper with u.v. light, which causes the whole paper to fluoresce except for the spots where a base is present, where absorption of the fluorescence occurs. This produces a series of dark spots against a bright background.

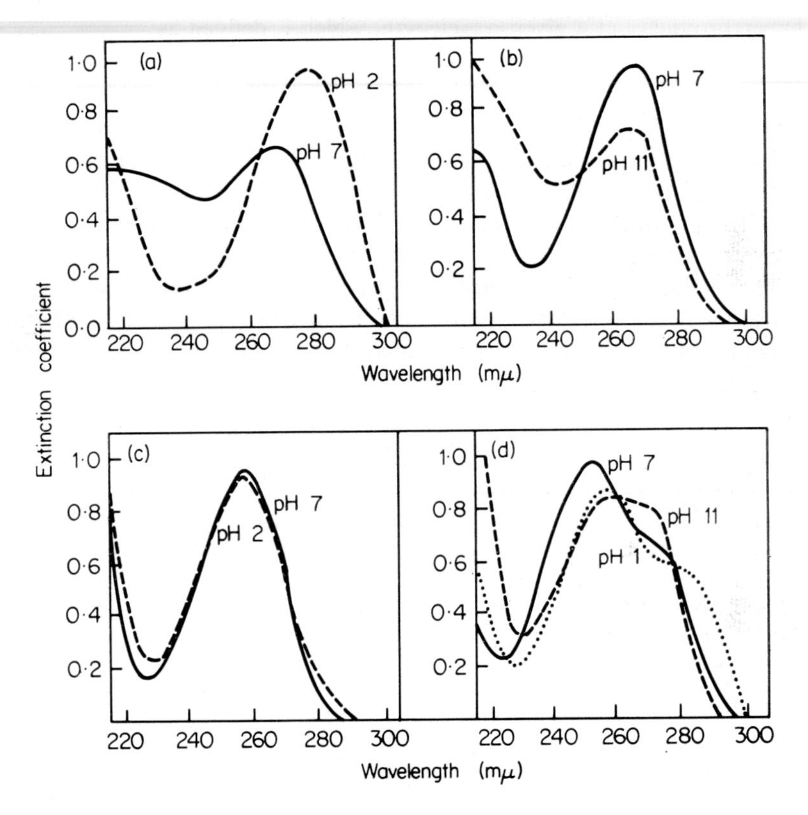

Figure 2.6. Ultraviolet absorption spectra of (a) cytidylic acid (5′-CMP), (b) uridylic acid (5′-UMP), (c) adenylic acid (5′-AMP) and (d) guanylic acid (5′-GMP)

When the bases are assembled into polynucleotides of ordered structure, the actual quantity of u.v. light absorbed is less than would be expected from the sum of the constituents. This property is termed hypochromism and the hypochromicity at any one wavelength ($H_{(\lambda)}$) can be defined by the following equation:

$$H_{(\text{at wavelength}\,\lambda)} = \frac{\sum^{c} A_{c(\lambda)}\,[\text{polymer}]}{\sum^{c} A_{c(\lambda)}\,[\text{monomers}]} < 1$$

where $H_{(\lambda)}$ is the hypochromicity at the wavelength λ and A_c the absorbancy of component c.

These data can also be presented in a slightly different form of h, the hyperchromicity at a given wavelength, which is defined as:

$$h_{(\lambda)} = (H^{-1}{}_{(\lambda)} - 1) = \frac{A_{(\lambda)}\,[\text{disordered form}]}{A_{(\lambda)}\,[\text{ordered form}]} - 1$$

The transition from an ordered structure to a disordered structure, or mixture of nucleotides, is accompanied by an increase in absorbancy, and it is the fractional increase in this value which is represented by the above equation.

The molar extinction coefficient, estimated on a mononucleotide or phosphorus basis, ε(P), allows the comparison of widely differing polynucleotides. Several properties of interest emerge from such a study, and a few examples are given in Table 2.1.

Table 2.1[a]

Polynucleotide	*Maximal hyperchromicity*	ε(P)
Poly r-A	0·54	9,500
Poly r-G	0·39	9,800
Poly (dG : dC)	0·42	7,400
Poly (dAdT : dAdT)	0·84	6,650
Ribosomal RNA	(thermal hyperchromicity) (0·21)	7,450
Transfer RNA	0·49	7,200
Calf thymus DNA (%GC = 41)	0·66	6,600
E. coli DNA (%GC = 50)	0·60	6,740

[a]After Mahler, Kilne and Mehrotra (1964), Inman and Baldwin (1964) and Inman (1964).

The hyperchromicities of DNA molecules always appear to be greater than those of RNA molecules (implying a more ordered structure in the former), and those of double-stranded polynucleotides greater than single-stranded ones, although even the latter can on occasion show a considerable residual hyperchromicity. Secondly, within a given polynucleotide sequence, the hyperchromicities vary directly with the content of the bases so that, for example, the maximal hyperchromicity of poly dAdT : dAdT = 0·84 whereas poly dG:dC = 0·42. As yet no one has put forward an exact explanation for this phenomenon, but as the optical transitions of interest are $\pi \rightarrow \pi^*$ and $n \rightarrow \pi^*$, the former being more important at 260 mμ, the normally expected transitions of the π-electrons could be being affected by interactions with other electrons within the stack of bases. This in turn would decrease the probability of any one particular excitation event within the macromolecule.

Hypochromism is important as a way of measuring the amount of order or disorder within a given macromolecule, particularly in T_m determinations, in which a given sample of DNA is gradually heated under controlled conditions. The absorbance at 260 mμ increases rapidly at a certain temperature, which is characteristic of the per cent GC in the molecules (Figure 2.7).

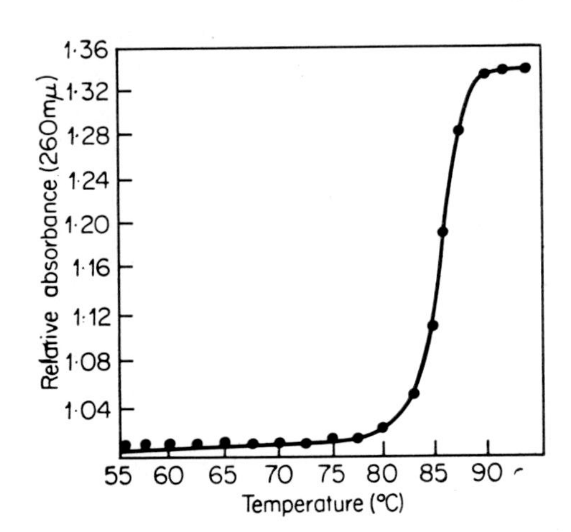

Figure 2.7. Thermal transition (T_m) of *B. subtilis* DNA (15 μg. of DNA in 0·15 M NaCl and 0·015 M trisodium citrate pH 7) as monitored by u.v. absorbancy

Hybridization

Hybrids of RNA and DNA can be isolated from *in vitro* RNA synthesizing systems or formed artificially by annealing individual RNA species with DNA.

The phenomenon of hybridization was first discovered by Marmur and Lane (1960) and Doty and coworkers (1960). They observed that the DNA helix could be reformed after thermal denaturation by cooling slowly to a low temperature, and that the renaturation was species-specific. Hall and Spiegelman (1961) subsequently succeeded in annealing RNA and DNA, and showed that T_2 virus-induced RNA synthesized in *E. coli* was complementary to T_2 DNA.

Several reviews are available which discuss the conditions for single-strand association, both in the case of DNA renaturation and RNA–DNA hybridization (Marmur and coworkers, 1963; Nygaard and Hall, 1964; Gillespie, 1968; Studier, 1969; McCarthy and Church, 1970).

The technique of hybridization is based upon the principle of base-pairing of complementary nucleic acid strands according to the Watson–Crick hydrogen-bonding hypothesis. Hence, RNA and its template or two separated DNA strands can hybridize to form double helices. There are two major requirements for successful hybridization. First, the double helix of the hybrid must be stable under the experimental conditions employed, the temperature must not exceed the T_m value and the ionic strength must be sufficient to overcome the repulsive forces which keep the charged phosphate groups apart. Second, the single-stranded molecules must be uncoiled in such a way that the base recognition sites are free for the pairing mechanism (Studier, 1969).

Hybridization is a second-order reaction, and the rate of formation of specific hybrids depends on the relative concentration of the complementary segments in the two single-stranded molecules taking part in the reaction, and hence on the degree of heterogeneity of the DNA. Viral DNA can therefore renature faster than bacterial DNA, which in turn renatures much more quickly than that of mammals or higher plants (Marmur and coworkers, 1963).

The two nucleic acid species can hybridize either in a homogeneous or heterogeneous phase system. In the first case, the two species are maintained in an aqueous salt environment at neutral pH and at a temperature close to the T_m of the higher melting species. The hybrids are subsequently formed by slowly annealing at lower temperatures, and can be easily recovered in an intact form for further characterization by methylated-albumin coated kieselguhr (MAK) chromatography (see Chapter 6), caesium chloride density-gradient centrifugation (see p. 56) or hydroxyapatite column chromatography (see Chapter 5), or tested for biological activity by transformation (see Chapter 1). Homogeneous phase reactions, however, necessitate the use of low DNA concentrations, low temperatures and low ionic strength.

More elegant hybridization techniques have now been developed involving heterogenous phase reactions, in which denatured DNA is fixed to a solid support. Bautz and Hall (1962) used DNA covalently linked to acetylated phosphocellulose, and passed RNA dissolved in a saline medium through the column at 55°C. A column of high molecular weight DNA trapped in agar gel was used by Bolton and McCarthy (1962, 1963) and by Walker and McLaren (1965). Homologous DNA or RNA was hybridized with DNA and the non-hybridizable fraction washed immediately from the column. The specific hybrid could then be eluted by a solution of lower ionic strength and at a higher temperature. The solid matrix prevents specific DNA–DNA renaturation, but the method has the disadvantage that the homologous DNA species must be sheared to low molecular weight so as not to become trapped within the agar matrix.

The technique most widely used at the present time is the nitrocellulose membrane filter technique (Gillespie, 1968). Nitrocellulose adsorbs single-stranded DNA and RNA hybridized to the DNA, but not free RNA. This type of filter was used by Nygaard and Hall (1963, 1964) to collect hybrids after annealing in solution. Gillespie and Spiegelman (1965), however, developed this method further. Denatured DNA was attached to nitrocellulose filters which were subsequently annealed with labelled RNA in aqueous salt solutions at 60°C for 6 h. The filter was washed and treated with RNAase to remove non-specifically bound RNA, and subsequently counted for radioactivity. The degree of adsorption of denatured DNA was directly proportional to the DNA chain-length and inversely proportional to the loading temperature. However, the nucleic acids must not be contaminated by protein as this tends to cause sticking.

The optimum conditions of hybridization must be determined for each RNA species under study. The degree of heterogeneity determines the ratio of RNA to DNA required for saturation of the DNA hybridization sites with RNA. The degree of saturation is usually expressed in terms of percentage hybridization and theoretically indicates the fraction of the genome coding for the particular RNA under study. Yankofsky and Spiegelman (1962, 1963) found that 0·1 to 0·2 per cent of the total *E. coli* genome coded for rRNA, and that non-ribosomal RNA did not compete for the same regions. Moreover, they observed that the two rRNA species of the ribosomal sub-units of *Bacillus megaterium* were each derived from unique genome segments, and estimated that the 23S cistron and 16S cistron occupied 0·2 and 0·1 per cent of the genome, respectively. The ability of *E. coli* tRNA to hybridize with DNA was also studied. Only 0·023 per cent of the genome was saturated and the tRNA–DNA complex was species-specific as shown by tests for competition by heterologous tRNA (Giacomoni and Spiegelman, 1962; Goodman and Rich, 1962).

Studies of the competitive effects between various RNA species for hybridization with DNA can be very useful in assessing the similarities or differences between RNA species. Schweizer, Mackechnie and Halvorson (1969) using *Saccharomyces cerevisiae* found values of 0·8, 1·6 and up to 0·08 per cent hybridization to nuclear DNA for 18S, 26S and 4S RNA, respectively. The three species did not compete with each other for the same sites to any significant extent, and would not hybridize to mitochondrial DNA. They estimated a value of 140 cistrons for the haploid genome for both 18S and 26S RNA. This agreed with the values for other eukaryotic systems, such as amphibians (Birnsteil and coworkers, 1968), plants (Chipchase and Birnsteil, 1963) and mammals (Attardi and coworkers, 1965). The percentage of DNA complementary to RNA in all the organisms varied between 0·1 and 0·3 per cent and was similar to that found in bacterial systems (Oishi and Sueoka, 1965; Doi and Igarashi, 1960).

The nitrocellulose filter technique has also been used to study the hybridization between rRNA and DNA isolated from the mitochondria of *Neurospora crassa* (Wood and Luck, 1969). The 23S and 19S rRNA's were found to be complementary to 6·1 and 2·8 per cent of the DNA, respectively, and could not be hybridized with nuclear DNA. However, the 28S and 18S rRNA of the cytoplasm were hybridized with nuclear DNA to an extent of 0·67 and 0·33 per cent, respectively.

Competitive hybridization, has also been used to show the existence of a long-lived mRNA in wheat embryo (Chen, Sarid and Katchalski, 1968). The total RNA labelled with ^{32}P between 24 and 48 h. of germination was complementary to 1·45 per cent of the genome, but rRNA and tRNA represented only 0·28 and 0·03 per cent, respectively. The authors therefore assumed that the difference (1·15 per cent) was the fraction of the genome complementary to mRNA synthesized during the particular period of germination.

Recently Nass and Buck (1970) have made some extremely elegant studies on mitochondrial tRNA from rat liver cells. They hybridized radioactive aminoacylated-tRNA's with mitochondrial DNA using the nitrocellulose filter technique. Competition experiments indicated that mitochondrial aminoacylated-tRNA competed to a greater extent with mitochondrial tyrosyl-, phenylalanyl-, seryl- or leucyl-tRNA for the hybridization sites on mitochondrial DNA than did the equivalent cytoplasmic aminoacyl-tRNA's. They observed no hybridization with rat liver nuclear DNA or *E. coli* DNA. It appeared that at least four mitochondrial tRNA's were potentially transcribed from mitochondrial

DNA, and that mitochondrial tRNA's differed in base sequence from their cytoplasmic counterparts. Hybridization studies of mitochondrial amino-acylated-tRNA's with isolated complementary heavy and light strands of mitochondrial DNA showed that mitochondrial leucyl- and phenylalanyl-tRNA hybridized only with heavy DNA strands, whereas tyrosyl- and leucyl-tRNA hybridized only with the light DNA strands. Some species of mitochondrial tRNA could therefore be transcribed *in vivo* from one strand, and others from the complementary mitochondrial DNA strand.

Sedimentation analysis

Macromolecules in solution do not move unidirectionally under the influence of the earth's gravitational force, and an external force, which is large enough to overcome the kinetic energy of the surrounding medium, must be applied. If the gravitational potential energy is raised above kT, where k = Boltzmann's constant and T = absolute temperature, then the particle begins to move through the medium in the direction of the force. This can be achieved by centrifugation. As the speed increases, at a certain point kT is exceeded, and the particles begin to move away from the axis of rotation. Unfortunately, centrifugal fields of force are not uniform, but vary with the distance from the axis of rotation. This means that if the angular speed is w radians/second, and the distance from the axis is r(cm.), then the linear acceleration (a) at that point is $a = rw^2$. According to the traditional definition of force, that is mass multiplied by acceleration, the force acting on a particle of mass m, in such a centrifugal field would be $ma = mw^2r$. However, if the density of the particle is equal to d, and the density of the solvent is equal to d_0, there will be a buoyant force acting on the particle in the opposite direction equal to md_0w^2r/d. Thus the net force becomes $mw^2r(1 - d_0/d)$.

Sedimentation velocity

The purpose of this method is to give one some idea of the mass of the particles being investigated. These data indicate, for example, which of the several species of RNA has been isolated, and whether or not any breakdown has taken place. The method is relatively simple but the interpretation of results must take into account several points which will be mentioned later. Theoretically, if a solution of particles is subjected to a centrifugal field, there will be a constant movement of the particles away from the axis of rotation. However, if the frictional coefficient is equal to f, and the velocity of the particles is equal to v, there will be a force of fv opposing this movement, together with the buoyant force of the solvent. At equilibrium, therefore, these three forces will balance, and the following will be obtained:

$$fv = ma\,(1 - d_0/d)$$

and if v is equal to the average velocity, then,

$$v = ma\,(1 - d_0/d)/f$$

or the mass of the particle could be deduced from

$$m = fv/a(1 - d_0/d)$$

In point of fact, this basically simple equation is difficult to use in practice because of the requirement of d, the density of the particle, which is difficult to measure. Since both the velocity and the acceleration can be easily measured for any particle, it has become common practice to categorize particles according to the ratio of v/a which is known as the sedimentation constant ($S = v/a$). The sedimentation constant is measured in the units (m./sec.)/(m./sec.^{2}), and, in memory of one of the earliest pioneers of ultracentrifugation, a value of S equal to 10^{-13} sec. is called 1 Svedberg. Thus particles are regularly known by their S values, e.g. 23S RNA.

There are several ways of observing the behaviour pattern of molecules sedimenting in such gravitational fields, and most sophisticated instruments offer a choice. When an initially homogeneous solution is exposed to such forces, the heavier molecules move rapidly away from the rotor axis, leaving the solvent and/or more slowly sedimenting material behind. This results in the formation of a boundary between the solvent at the 'top' of the cell and solution which is moving to the 'bottom'. The movement of this boundary along the length of the cell is followed, and from a plot of time against distance moved, a value of the rate of movement is obtained.

The nucleic acid sample is placed in a cell, the upper and lower surfaces of which are made of quartz. U.v. light is passed through the cell and the moving boundary is photographed at various time intervals and developed (Schachman, 1959; Chervanka, 1969). A pattern forms in which the lighter region towards the bottom of the cell becomes progressively smaller, while the darker region at the top of the cell lengthens, the line between these two regions forming the boundary (Figure 2.8).

Alternatively a system termed Schlieren optics can be employed which uses the fact that an incident wave front of light passing through a solution containing two regions of unequal concentration will no longer be planar on emergence. The wave front is retarded by the more concentrated solution, which has a higher refractive index than the solvent. The emergent wave front therefore shows the concentration gradient over the boundary region and remains planar (although retarded) in the solution and solvent. The distortion over the boundary region which changes with time can be observed in several ways. The observed sedimentation coefficient can then be calculated from the equation

$$S_{obs} = 1/rw^2 \,.\, dx/dt$$

where x is the distance from the rotor axis to the boundary and t is time.

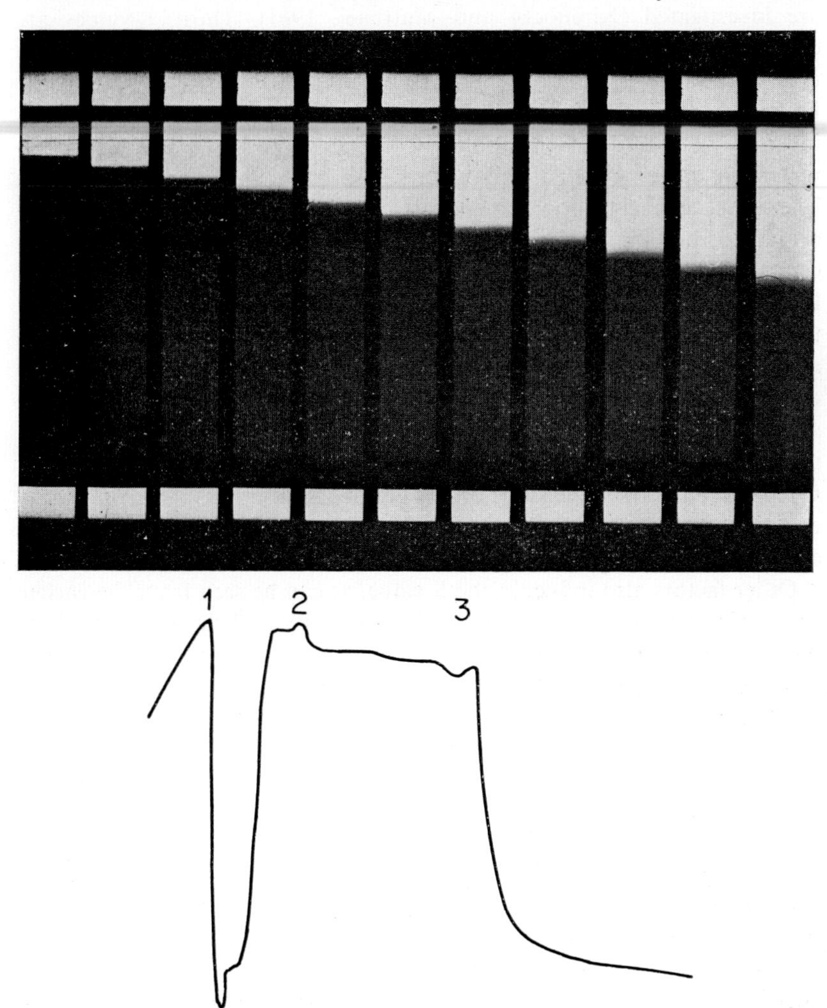

Figure 2.8. Ultracentrifuge sedimentation pattern of *B. subtilis* DNA (30 μg./ml. of DNA in 0·15 M NaCl and 0·015 M trisodium citrate, pH 7). Photographs were taken using u.v. optics every 4 min. after reaching speed of 35,600 r.p.m. at 20°C. Photographs were scanned using Beckman Analytrol film densitometer, (1) is the reference mark, (2) is the meniscus and (3) is the DNA boundary. (After Ayad (1964))

The boundary observed is influenced by several factors: First, simple diffusion of the solute back into the solvent can broaden the boundary. Secondly, the polydispersity of the solute also causes boundary broadening as the faster and slower moving components begin to separate. The latter factor is particularly noticeable when degraded nucleic acids (Figure 2.9) are investigated (Nirenberg and Matthaei, 1961). Third, S-values are concentration-dependent. Since the solute molecules on the trailing edge of a boundary will be more dilute than those on the leading edge of the boundary, their S-values will be slightly different and, since S varies inversely with the concentration c (see below), these trailing molecules will move faster, reach the boundary and cause boundary sharpening. The exact concentration dependence of S has not been satisfactorily evaluated, but the equation $S = S^0/1 + kc$ is often applied. This concentration dependence is particularly noticeable with DNA where, at higher concentrations, the S-value can fall to a much lower value than that observed at very small concentrations, but the effect is not so noticeable with spherical molecules. Quoted S-values must be defined in terms of the concentration at which sedimentation was carried out or, more accurately, a series of S-values can be measured at differing concentrations, and the results plotted in the form of 1/S against c. The constant, k, can be found from the slope of this graph and inserted into the previous equation, or the intercept of the line at $c = 0$ obtained by extrapolation, and the value of S^0 quoted.

Other factors also influence the S-value, as can be seen from the earlier equations. These all concern the solvent density and depend, therefore, on the nature of the solvent, its ionic strength and the temperature (Vinograd and coworkers, 1963; Studier, 1965). Corrections for these must also be taken into account, and the final S-values quoted should be converted to S^0_{20w} which is defined as the S-value obtained for this component at zero concentration in pure water at 20°C (Studier, 1965).

Density gradients

In the experiments described above, a homogeneous solution containing the solute is subjected to a centrifugal force and the solute separates in a zone of higher concentration. This system is used to investigate those physical properties which are dependent on S-values (i.e. molecular weight), and to obtain some idea of molecular heterogeneity. There are, however, practical limitations to the amount of material it is possible to use with accuracy, and the analytical ultracentrifuge cannot be used to recover individual components which have been separated or partially separated in the cell. To overcome these difficulties, and others caused by convective disturbances and solute–solute interactions, alternative ways have been used in which the sample either bands, or moves through a continuous gradient of a denser material, under the influence of a centrifugal field of force. The manner in which it does this characterizes its

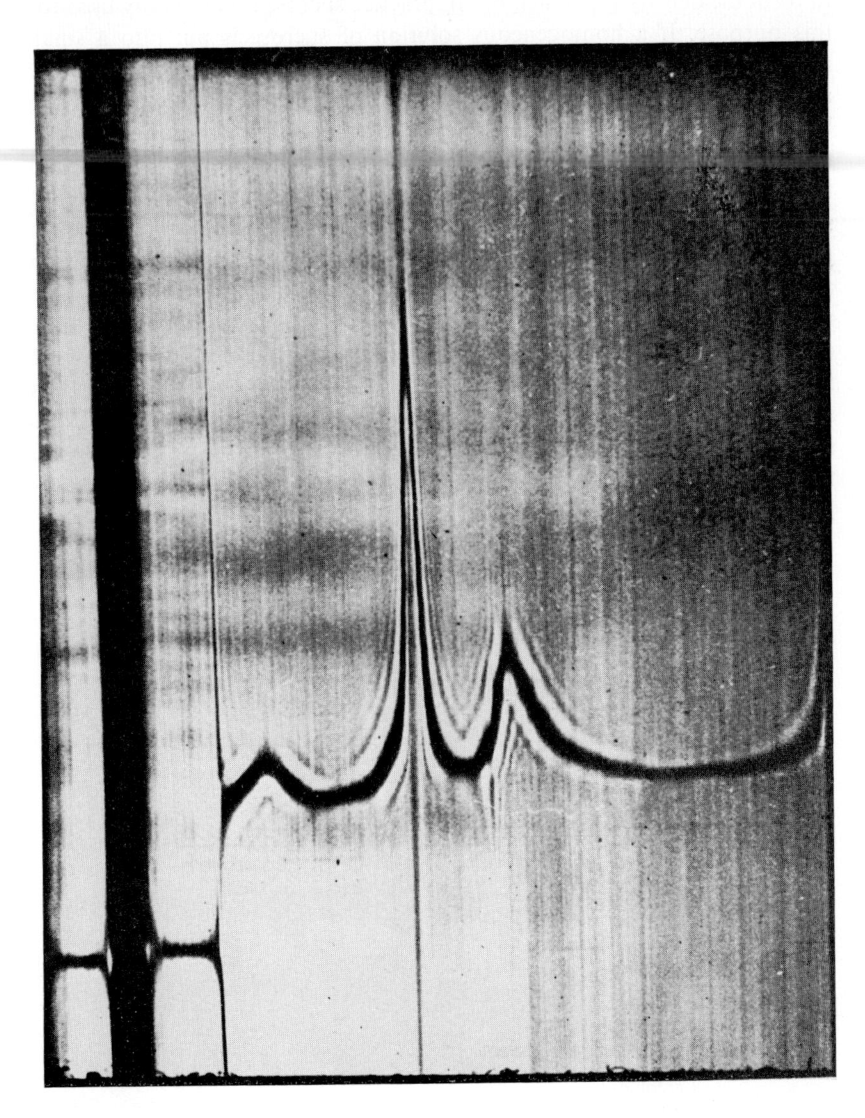

Figure 2.9. The sedimentation characteristics of *E. coli* ribosomal RNA using model E spinco analytical ultracentrifuge equipped with Schlieren optics, showing the 23S, 16S and 4S sub-units. (After Nirenberg and Matthaei (1961))

molecules. The two approaches are known as rate sedimentation and equilibrium sedimentation.

Theoretically, any medium sized molecular species, which forms a true solution in buffer over a wide range of concentration, could be used to form the density gradients, but in practice sucrose is commonly used for this purpose. If a homogeneous solution of sucrose is put into a small round-bottomed centrifuge tube, and subjected to centrifugal force for a sufficient length of time, the sucrose molecules tend to move to the bottom of the tube. This tendency is eventually balanced by diffusive forces pushing the molecules back towards the surface. At equilibrium there is no net movement of sucrose and analysis of the quantities of sucrose at various points in the tube shows that a linear concentration gradient has been formed. This is the starting point for the sedimentation analysis, but it is obviously impractical to run an ultracentrifuge for what might be several days to form a linear sucrose gradient prior to the start of an experiment. Several ingenious ways have therefore been devised for preforming sucrose gradients in the tubes. A simplified example is shown in Figure 2.10. One chamber of the device is filled with a dilute solution of buffered sucrose, which is continually stirred. The outlet to the second chamber containing the concentrated buffered sucrose is

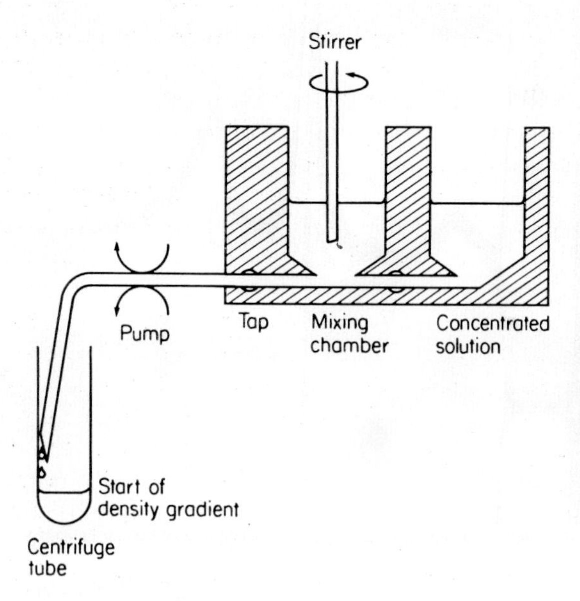

Figure 2.10. Diagrammatic representation of the apparatus used to produce a linear sucrose gradient

opened, as is the outlet to the centrifuge tube, and liquid is pumped, or allowed to drain, into the tube from the mixing chamber. As liquid is withdrawn from the mixing vessel, it is partially replaced by the concentrated solution draining into it, and the levels of liquid in the two chambers fall at the same rate. Theoretically the greater density of the concentrated sucrose solution should cause this liquid to flow into the mixing chamber with a greater velocity than that with which the mixed liquid is withdrawn. However, in practice, with small amounts of liquid, this effect does not appear to be very pronounced, and subsequent analysis of the density gradients produced show them to be both linear and reproducible. The gradients usually employed vary linearly between 5 and 20 per cent sucrose concentration, but many variations have been used, and it is not uncommon to find stepped gradients in which one solution of sucrose is simply layered on top of a more concentrated solution without mixing. During centrifugation, the 'steps' become smoother and eventually disappear.

In order to carry out the sedimentation experiment, a sample of the material under investigation is layered on to the top of the preformed gradient in a centrifuge tube, which is then placed in a preparative centrifuge head of either the angle or swing-out bucket-type. Sucrose gradients can also be formed and used in the analytical ultracentrifuge, but this is less common. A centrifugal force is then applied for a sufficient length of time to ensure that the molecules have moved into the gradient. The distance travelled is almost proportional to the time of centrifugation and the square of the rotor speed, so that for molecules with an estimated S-value in the region of 40S–50S, 3 to 4 h. at 20,000 r.p.m. is usually sufficient. The tube can then be treated in one of two ways depending on the aim of the experiment, If the experiment is designed simply to determine approximate S-values, the sedimenting material can be continuously monitored using the u.v. optics. However, if the experiment is being performed on a semi-preparative basis, it is necessary to recover the individual components from their zones in the tube without causing too much disturbance, and without too much back-mixing. The ingenuity which has gone into devising apparatus for this purpose is almost as great as that which has gone into devising gradient makers. However, the standard equipment used is shown in Figure 2.11. The bottom of a firmly held tube is punctured with a hollow needle, and the liquid in the tube either pumped out, or forced out by pumping oil on to the surface of the liquid, Drop fractions can then be collected either individually or in required groups, depending on the resolution required. The presence of nucleic acids in any particular fraction can be monitored by u.v. absorbance or more specifically by a chemical assay. As a radioactive label is frequently incorporated into the molecule, the drops of liquid can also be monitored for radioactive decay.

Stewart (1969) has used zone sedimentation to investigate the heterogeneity of *B. subtilis* DNA when purified by a standard procedure. A sample of the DNA was sedimented through a sucrose gradient in the manner described (3 h. at 35,000 r.p.m.), and a broad smooth peak pro-

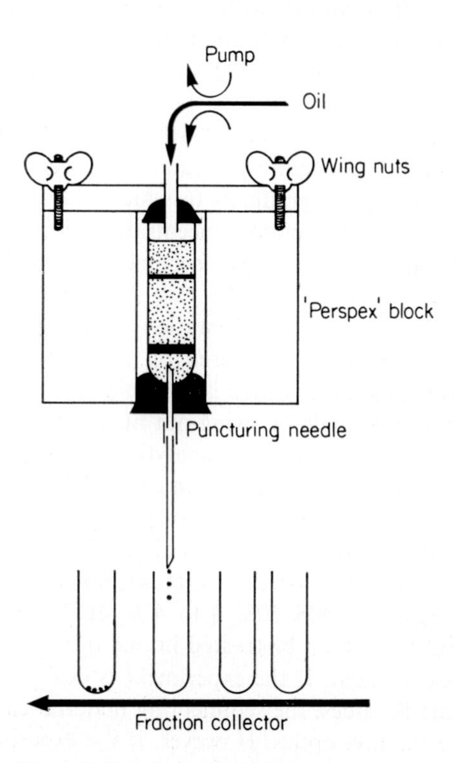

Figure 2.11. Diagrammatic representation of a device used for the collection of various fractions separated in preparative CsCl density-gradient centrifugation

duced. Fractions were then taken from each edge of this peak, and resedimented in separate sucrose gradients, the results of which are shown in Figure 2.12. The fractions sedimenting on the leading edge of the original peak sedimented to a distance down the tube equivalent to the original position, and similarly with the slower sedimenting fractions. The original peak, therefore, represented a true heterogeneity of molecular species in the first fractionation, in which the larger molecules travelled further down the gradient than did the shorter molecules. The presence of DNA in the fractions was detected by assaying for incorporated tritium, and the fractions were also tested for transforming ability. The results appeared to demonstrate that, within any particular preparation of DNA

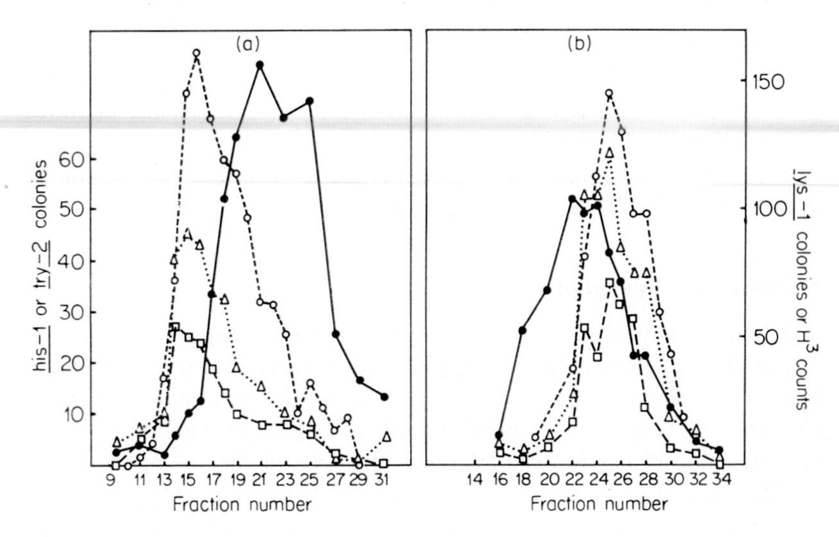

Figure 2.12. Sucrose gradients of long and short molecules. Samples (0·1 ml.) of solutions I and II were mixed with 5 μl. of approximately 10^{-4} M SB-1012 DNA (lys-1$^+$), which was used as a standard. These solutions were sedimented in separate sucrose gradients. A 20 μl. amount of each fraction was assayed for radioactivity, and 50 μl. was assayed by transformation of SB-1017. Each transformation mixture contained 1·0 ml., and the graphs show the number of colonies per 0·2 ml. of this mixture. (a) Solution I, long molecules; (b) solution II, short molecules. Symbols: ●, lys-1$^+$ (standard) colonies; □, try-2$^+$ colonies; △, his-1$^+$ colonies; ○, ^{3}H counts per 10 min. in (a); ^{3}H counts per min. in (b). (After Stewart (1969))

from *B. subtilis*, individual transforming markers are carried on DNA molecules which are variable in size. This might suggest that there is random breakage of the bacterial chromosome during isolation, but various groups of workers do not agree entirely with this point.

The experiments of Stewart (1969) show that heterogeneity of molecular size, particularly in DNA, causes a broad peak in sedimentation profiles, which is a true reflection of the amount of material at any point having a discrete molecular dimension. This is similar to the results obtained in the analytical ultracentrifuge, where the boundary was diffuse due to the heterogeneity of molecular species. However, these methods are also used to examine highly pure preparations which may contain two molecular species of differing molecular weight. A good example is the separation obtained between 23S and 16S RNA from bacterial ribosomes using either of the two methods. It is possible to calculate the S-value of a component from the distance it has moved into the gradient using complex equations involving viscosity, density and partial specific volume. However, a more convenient method is to use a marker molecule, with a known S-value, as a standard. Stewart (1969) used *B. subtilis*, SB-1012 DNA as a standard in his sucrose gradient analysis. The S-value of the unknown can then be calculated from the very simple relationship: S_{20w} (unknown) $= S_{20w}$ (known) $\times$ (distance travelled from meniscus by unknown/distance travelled from meniscus by known).

Sedimentation equilibrium

A homogeneous solution of sucrose would, if centrifuged for a sufficient length of time, form a density gradient, but with sucrose this takes a long time. If, instead of sucrose, a heavy-metal salt is used, such as CsCl or Cs_2SO_4, the formation of density gradients by centrifugation is not so impractical and, if large enough concentrations of the salt are used, the resulting range of densities is sufficiently wide to determine the density of the macromolecules under investigation. This means that a sample sedimenting through a CsCl density gradient, formed by ultracentrifuging a concentrated solution of the salt (about 7·7 M), will not travel to the bottom of the tube but will, upon reaching the zone of CsCl which has the same density as itself, 'float' in this layer, move no further, and a band will form at the point where the sample has the same buoyant density as the surrounding CsCl. This technique, therefore, offers the research worker a relatively simple and very practical way of determining the density of a given molecular species, as shown in Figure 2.13 (Sueoka, Marmur and Doty, 1959).

Density-gradient centrifugation is a simpler method than sedimentation velocity in sucrose gradients. Since the density gradient is formed during centrifugation it is only necessary to mix the DNA sample (in 0·02 M tris buffer, pH 8·5) with a suitable amount of CsCl known to give the required range of final densities in the tube, and to centrifuge at the required speed for sufficient time to allow a close approach to equilibrium. This procedure can be carried out in an analytical ultracentrifuge cell or in a centrifuge tube as described above, and the band of sample detected by any appropriate method. The density of an unknown can be calculated from a comparison of its position in the cell with that of a sample of known density (Figure 2.14), e.g. the DNA of *E. coli* B (50 per cent GC)

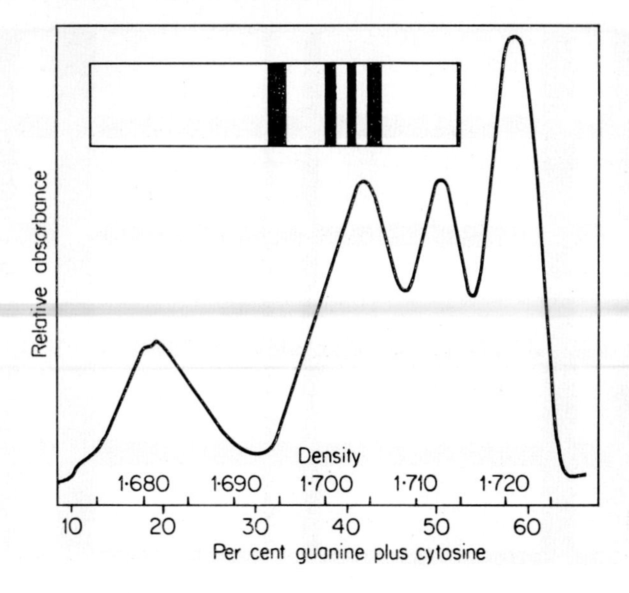

Figure 2.13. Resolution of polydeoxyadenine-thymine and *Pneumococcus*, *E. coli* and *Serratia marcescens* deoxyribonucleic acids by density-gradient centrifugation. The photograph was taken after centrifugation at 44,770 r.p.m. for 24 h. The tracing was taken with a microdensitometer in the region of the four bands. (The composition scale does not apply to the adenine-thymine deoxyribonucleic acid). (After Sueoka, Marmur and Doty (1959))

is one possible standard and has a density ρ equal to 1·710g./cm.^{3} (Schildkraut, Marmur and Doty, 1962). In semi-preparative work, the density of the CsCl solution at any point can be determined from the relationship between the density of the solution and its refractive index (Ifft, Voet and Vinograd, 1961).

$$\rho^{25^\circ C} = 10{\cdot}8061\, n_D{}^{25^\circ C} - 13{\cdot}4974$$

where $\rho^{25^\circ C}$ is the density at 25°C and $n_D{}^{25^\circ C}$ is the refractive index (sodium light) at 25°C. Thus if the refractive indices of the fractions on each side of the band of material are measured, it is possible to construct a straight line of refractive index against position (assuming a linear gradient has been formed), and hence deduce the exact refractive index that the centre of the band should have if there were no sample present. From this value, the density of the CsCl, and hence the sample, at that point can be calculated.

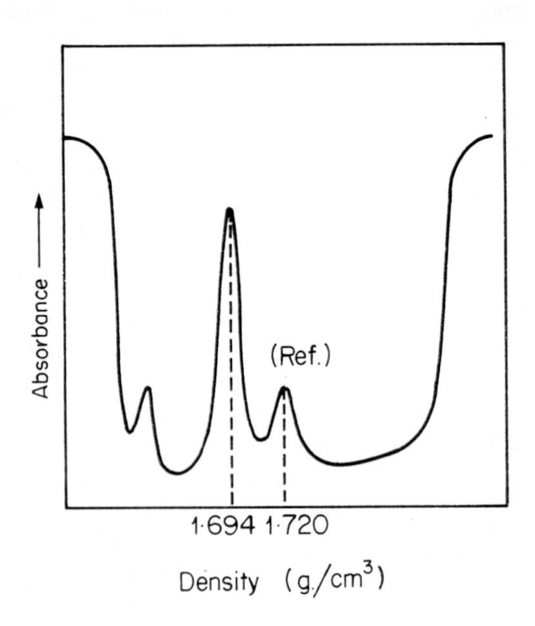

Figure 2.14. Mouse lymphoma DNA and heat-denatured *B. subtilis* DNA (as a reference) banded in CsCl density gradient after 24 h. at 44,770 r.p.m. (After Ayad (1968))

This system is applicable to DNA in its native double-stranded form, but single-stranded polynucleotides, such as denatured DNA and RNA, bind Cs^+ ions and the buoyant densities of such molecules in CsCl gradients are greater than their actual densities. This is not, however, a disadvantage, since it allows the separation of single-stranded DNA, which has an increase in buoyant density of about 0·016 g./cm.^{3}, from native DNA (Schildkraut, Marmur and Doty, 1962). RNA with a buoyant density in the region of 2·0 may also be separated from both.

The density of double-stranded DNA is an almost linear function of its guanine and cytosine (G + C) content so that, $\rho = 1{\cdot}660 + 0{\cdot}098$ (per cent GC) (where per cent GC is expressed as the mole fraction of guanine plus the mole fraction of cytosine) (Sueoka, Marmur and Doty, 1959; Schildkraut, Marmur and Doty, 1962). The presence of isotopes which have a higher atomic weight than the natural atoms in DNA, also causes a shift in the buoyant density to higher values. The classical example of this is the experiment of Messelson and Stahl (1958) which provided evidence in support of the semi-conservative model of DNA replication.

Applications of caesium chloride density-gradient centrifugation

The identification of extrachromosomal (satellite) drug resistance factors (R-factors) in bacteria as DNA, and a possible mode of isolation, were first recognized when hosts such as *Proteus mirabilis* and *Serratia marcescens* (which possess DNA with average base composition different from that of the hosts which contain the R-factor), could accept a transmissible genetic element during conjugation, and maintain it in subsequent generations.

The separation of satellite DNA from chromosomal DNA can be achieved by equilibrium density-gradient centrifugation, if the base composition (and hence density) of the R-factor is sufficiently distinct from the DNA of the host bacterium. The percentage G + C content of the R-factor can then be estimated from the determined density. For example, R-factor DNA of 58 per cent G + C content could be identified using *E. coli* (DNA, 50 to 52 per cent G + C) or *Proteus* (DNA, 38 per cent G + C) but not *Aerobacter* (DNA, 56 to 58 per cent G + C) as the recipient bacterium. This technique was first used to study the DNA of the extrachromasomal sex factor F merogenotes in *P. mirabilis* (Marmur and coworkers, 1961; Falkow and coworkers, 1964), and subsequently to study the DNA of an R-factor (Falkow and coworkers, 1966; Rownd, Nakaya and Nakamura, 1966). Using *Proteus* as the recipient host, both factors were found to have a buoyant density characteristic of *E. coli* ($\rho = 1{\cdot}710$ gm./cm.^{3}, G + C = 50 per cent). Certain R-factors were found to give rise to an additional satellite DNA band at a density of 1·718 g./cm.³ corresponding to a G + C content of 58 per cent. The R-factor DNA of the two satellite bands accounted for 7 per cent of the total chromosomal DNA (Falkow and coworkers, 1966). On the other hand, the R-factor NRI gave only one band at a density of 1·718 g./cm.³ (G + C content of 58 per cent), but represented 12 per cent of the total extrachromosomal DNA (Rownd and coworkers, 1966).

The presence of more than one density band associated with a particular R-factor was thought by Rownd and coworkers (1966) to result from a shearing of the DNA molecules during the isolation procedure. However, Nisioka, Mitani and Clowes (1969) separated R_{222} DNA from *Proteus* DNA by CsCl density-gradient centrifugation, the R-factor DNA consisting of three distinct density bands (Figure 2.15). Electron microscopy

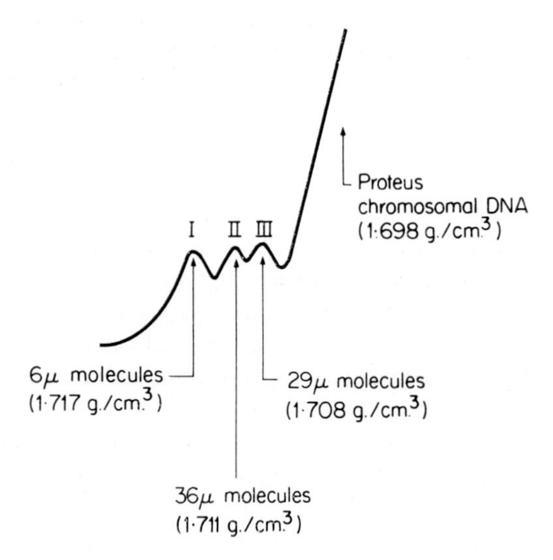

Figure 2.15. Schematic diagram of density-gradient separation and contour length of R-factor DNA molecular species. (After Nisioka, Mitani and Clowes (1969))

indicated that 10 per cent of the DNA in the bands was in the form of closed-circular DNA molecules, and that their average contour length depended on the density of the fraction: fractions I, II and III had densities of 1·717, 1·711 and 1·708 g./cm.³ and contained circular molecules with an average contour length of 6·4, 35·8 and 28·5 μ, respectively. Bands I and III corresponded to components with 58 per cent and 50 per cent G + C content, respectively. The intermediate fraction II was found to contain circular molecules with a contour length and density that could have resulted from a recombination of molecules I and III. The demonstration of three types of closed circular DNA indicated that Rownd and coworkers (1966) were incorrect in their assumption that the presence of two bands, corresponding to I and III, resulted from DNA degradation.

The closed-circular nature of extrachromosomal DNA can be utilized to effect its preliminary separation from linear or nicked-circular DNA by methods which do not necessitate bacterial conjugation. One method involves the phenomenon of dye intercalation (Waring, 1968, 1970) and another the selective denaturation and removal of denatured DNA by nitrocellulose.

The dye ethidium bromide is known to intercalate between the base-pairs of the DNA double helix resulting in a local unwinding. The restricted configuration of closed-circular DNA, however, allows only a limited amount of uncoiling in the presence of ethidium bromide. Uncoiling is manifested by an increase in the molecular length and hence a reduced buoyant density. Closed-circular DNA, which retains less ethidium bromide per unit length than nicked-circular DNA, has therefore a greater buoyant density. The two types of DNA can therefore be easily separated pycnographically by CsCl density-gradient centrifugation. The changes in sedimentation which also occur when DNA is uncoiled have been used by Crawford and Waring (1967) and Waring (1970) to estimate the number of super coils present in closed-circular DNA.

Ethidium bromide, followed by CsCl density-gradient centrifugation has been used to detect closed-circular DNA in the mitochondria of HeLa cells (Radloff, Bauer and Vinograd, 1967) and to characterize this DNA in the form of catenated oligomers (Hudson and Vinograd, 1967). Bauer and Vinograd (1968, 1970 a, b) used this method to show the differences between open and closed forms of DNA's isolated from polyoma virus (SV40), *Micrococcus lysodeikticus*, *Clostridium perfringens*, Ø-X174 and crab d (A — T). The method has also been used successfully to study P_1 prophage in *E. coli* (Ikeda and Tomizawa, 1968) and the nature of the closed-circular DNA associated with colicin factors E_1, E_2 and E_3 in *E. coli* (Bazaral and Helinski, 1968 a, b).

Closed-circular DNA also behaves differently from linear and nicked-circular DNA towards alkali denaturation. Closed-circular DNA denatured by alkali, subsequently renatures to its native state upon neutralization, whereas nicked-circular and linear DNA are irreversibly denatured. Methods which distinguish between native and denatured DNA can therefore be used to isolate the alkali-resistant closed-circular DNA. Selective alkali denaturation followed by CsCl density-gradient centrifugation, has therefore been successfully used by Rush and coworkers (1969) to examine penicillinase plasmid DNA from *Staphylococcus aureus* and heterogeneous cryptic plasmid DNA from *Shigella dysentariae*, Y_6R.

Cohen and Miller (1969) utilized the ability of closed-circular DNA to resist alkali denaturation, to isolate large amounts of closed-circular R-factor NRI DNA from *E. coli* during the log phase of growth. They used nitrocellulose preferentially to bind denatured DNA. The cells were lysed, exposed to alkaline pH and subsequently neutralized. A large proportion of the DNA was denatured and bound to the nitrocellulose, whereas the alkali-resistant closed-circular DNA remained in the supernatant, and was fractionated in a CsCl density gradient. These closed-circular DNA molecules were shown by electron microscopy to be 32 μ in circumference and to have a molecular weight of 63 to 65 $\times$ 10^6. There was also a minor circular DNA component (representing three to six per cent of the total circular molecules) with an average contour length of 5·5 μ. Multiple molecular species of R-factor may occur in *E. coli*, as was found with *P. mirabilis*.

Nitrocellulose has been used effectively to separate closed circular native DNA from denatured material. However, other fractionation techniques, such as hydroxyapatite and polylysine kieselguhr chromatography (see Chapters 5 and 6), may be just as effective, and can be used to isolate circular and nicked-circular DNA in greater yield.

Preparative CsCl density-gradient centrifugation has been used in conjunction with the ^{15}N density labelling technique devised by Meselson and Stahl (1958) to study the replication of R-factor NRI in *P. mirabilis* (Rownd, 1969). The appearance of $^{15}N/^{14}N$ DNA hybrid and $^{15}N/^{15}N$ heavy DNA, synthesized after transferring the culture from ^{14}N- to ^{15}N-containing medium, was followed by CsCl density-gradient centrifugation. Chromosomal DNA also formed hybrid molecules during the first generation of growth in ^{15}N-containing medium, and overlapped with the R-factor bands. However, the amount of DNA in the various regions of the gradient could be estimated by using a curve resolving machine. The relative amounts of DNA in the light, hybrid and heavy R-factor regions indicated that the R-factor NRI was replicated by a random mechanism. A similar conclusion was reached by Bazaral and Helinski (1970) while studying the replication of Col E_1 in *E. coli*.

Preparative CsCl density-gradient centrifugation has also been used to demonstrate 'repair synthesis' in mammalian cells, after treatment with X-rays (Painter and Cleaver, 1967; Ayad and Fox, 1969, 1970), u.v. irradiation (Cleaver and Painter, 1968), nitrogen mustard (Roberts, Crathorn and Brent, 1968) and methylmethane sulphonate (Ayad, Fox and Fox, 1969). Furthermore, the mechanism of uptake and integration of exogenous DNA has been studied, using CsCl gradients, in mammalian cells (Ayad and Fox, 1968, 1969; Robins and Taylor, 1968) and in bacterial cells (Bodmer and Ganesan, 1964; Pene and Romig, 1964; Ayad and Barker, 1969).

Summary

The techniques described have mainly an analytical value and are used to characterize the size, shape, physical state (i.e. double- or single-stranded, linear, circular or supercoiled) and the base composition of nucleic acid mixtures or fractions obtained by column chromatography. Conversely, knowledge of the physical properties prior to fractionation can be used to choose a particular preparative procedure. Certain analytical techniques (notably hybridization and CsCl density-gradient centrifugation) may also be used preparatively as a preliminary or main stage in fractionation.

CHAPTER 3

Ion exchange chromatography

Introduction

The mathematical treatment of ion-exchange chromatography is extremely complex, and the reader should refer to the original literature for precise details. In brief, exchange between oppositely charged ionic species occurs. If R^+A^- represents a positively charged polymer resin (immobile matrix) neutralized by a small negatively charged atom or molecule and P^-B^+ is the negatively charged nucleic acid macromolecule neutralized by small positively charged ions, then when the two species come into contact, exchange is possible as follows:

$$R^+A^- + P^-B^+ \rightleftharpoons R^+P^- + B^+A^-$$

At the top of the column matrix the species R^+A^- and P^-B^+ will be in high concentration and, by the law of mass action, the reaction will go to the right. This results in the 'trapping' of the macromolecule P^- on to the matrix R^+ and B^+A^- passes through the column. The column can then be 'developed' by increasing the concentration of B^+A^- (or similar ionic species) so that the reaction moves to the left, and the released components can pass further down the column. This introduces the concept of 'theoretical plates', in which the column is considered to be made up of a large number of segments or 'plates', each being the thickness of the constituent matrix beads. Each of the segments in turn achieves complete equilibrium, in which the solute is distributed between the matrix and the solution in a constant proportion (the distribution coefficient), D, i.e. the ratio of the amount of solute in the matrix to that in the solution in any given plate. The solute, originally a thin band at the top of the column, diffuses into a broader band having a maximum concentration in a certain segment, with smaller concentrations following and preceding it. If v is the volume of solution retained by the column (the column volume), and V is the volume of eluting buffer required for the solute to migrate from the top of the column to the bottom (elution volume), the following relationship can be deduced:

$$D = V/v$$

This relationship is obeyed in practice but the bands are often broader than the theory predicts. Since polynucleotide molecules are polyanions, the type of ion-exchanger required to trap them is of the anion-exchanger type.

Anion exchangers

Wood cellulose was used as the supporting matrix for anion exchangers because of its hydrophilic nature and enormous surface area. Moreover simple modifications to the reactions extensively used in the industrial manufacture of cellulose products, provided a means of attaching a variety of ionizable groups to the cellulose matrix. There were, however, certain limitations. Hydrogen bonding between the hydroxyl groups on the cellulose molecule is responsible for its insolubility, and a small degree of substitution interferes with this hydrogen bonding, causing the compound to swell in water.

ECTEOLA-cellulose (the first compound prepared by Sober and Peterson (1954)) is now obsolete. However, the use of this compound did help in the understanding of certain problems of nucleic acid fractionation. One serious disadvantage was the degradation of high molecular weight rRNA (Goldthwait and Kerr, 1962) during elution. Kit (1960), however, obtained reproducible results for chromatography of various DNA preparations if precautions to minimize degradative and denaturation processes were enforced. The disadvantages of degradation, however, did not prevent the method being successfully used for small tRNA molecules, and Goldthwait and Starr (1960) were able to separate an RNA fraction from the bulk of the RNA soluble in M NaCl, which stimulated the incorporation of ^{14}C-leucine. ECTEOLA-substituted cellulose has now, however, been largely superseded by diethylamino-ethyl-substituted (DEAE-) matrices.

DEAE-cellulose and DEAE-Sephadex column chromatography

DEAE-cellulose columns were used by Otaka, Mitsui and Osawa (1962) to separate DNA, tRNA and several classes of rRNA. Moreover, rRNA precursor was identified by labelling the RNA as it was formed on the DNA strands. DEAE-cellulose fractionation was also used: (*a*) to isolate a DNA–RNA hybrid which accumulated when the microbial cells were incubated with chloramphenicol, and (*b*) to trace the conversion of rRNA precursor into ribosomes.

The cellulose fibres are themselves weak ion-exchangers and secondary binding forces between cellulose and purine and pyrimidine derivatives can be readily shown. For example, the R_f values of nucleosides, which carry no electrical charge, on paper (cellulose) chromatograms are dT = 0·79, dC = 0·77, dA = 0·53 and dG = 0·60 in water at 22°C. If no interaction had occurred between the nucleosides and the cellulose, the compounds would have moved with the solvent front. These secondary forces are more important when considering the separation of short-chain polynucleotides and cause serious overlapping of adjacent peaks.

The use of urea in DEAE-cellulose chromatography

Tomlinson and Tener (1963a) used DEAE-cellulose columns to investigate the oligomers produced by DNase digestion of salmon testis DNA. The digestion product was eluted from the column using a linear gradient of ammonium carbonate (pH 8·4). About five regions could be distinguished, but only the first two were completely resolved (Figure 3.1a). The inclusion of 4 M urea (Figure 3.1b) and 8 M urea (Figure 3.1c) in the eluting buffers considerably improved the elution profile, and in the latter case 7–8 peaks could be seen. These successive peaks (Figure 3.1) were characterized by: (*a*) the order of elution from the column, (*b*) the ratio of phosphomonoesterase-sensitive phosphorus/total phosphorus, (*c*) the decreasing E/P (extinction/phosphorus) value with increasing size and (*d*) the ratio of free 5′- or 3′-hydroxyl to total nucleotide content obtained by treating the material with phosphomonoesterase, followed by snake venom or spleen phosphodiesterase. This treatment gave, respectively, mononucleotides (tubes 31–40), dinucleotides (tubes 43–54), trinucleotides (tubes 55–69) and longer polynucleotides in subsequent tubes. Each peak differed from the adjacent peaks by one negative charge.

The presence of urea allows mixtures of deoxyribonucleotides and oligomers to be separated according to the number of net negative charges on each polymer. Before this procedure was adopted to overcome secondary (possibly hydrogen-bonding) forces, Bell, Tomlinson and Tener (1963) had not been able to resolve completely the digestion products of yeast tRNA on DEAE-cellulose. The use of 7 M urea results in a clear separation of nucleotides with one or two phosphate groups, from those with no phosphate attached, and utilizing this property, Tomlinson and Tener (1963b) devised a general method for the isolation of the end group of any nucleic acid species. In this method, the nucleic acids are attacked by specific nucleases in such a way that the terminal group (nucleotide) has either two phosphates or no phosphate. The passage of this hydrolysate through a DEAE-cellulose column in 7 M urea separates the nucleotides from the nucleosides, since the latter are not held on the column but are

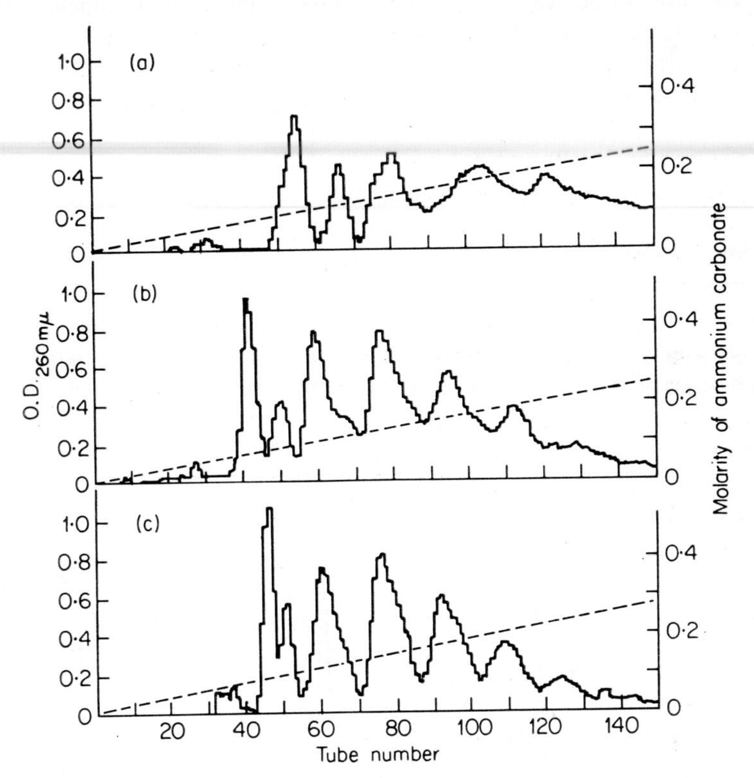

Figure 3.1. The separation of components of DNase I digest (550 O.D. units at 260 mμ) of salmon testis DNA on a DEAE-cellulose (carbonate) column (35 × 2·4 cm.); eluting solution, a linear gradient of ammonium carbonate (pH 8·4) containing (a) no urea, (b) 4 M urea and (c) 8 M urea. Fraction size 18 ml. (After Tomlinson and Tener (1963a) *Biochem.* **2,** 697. Copyright 1963 by the American Chemical Society. Reprinted by permission of the copyright holder)

immediately eluted. The monomers can then be eluted in the usual way, or released in a stepwise fashion and treated with phosphomonoesterase. Thus the end group, if it previously had two phosphates will still contain one phosphate group, whereas the monophosphates will have no phosphate attached. When the nucelotides are passed through the DEAE-cellulose column for the second time, and eluted with 7 M urea, only the end-group nucleotide which still has a negative charge will be retained. In this way, the end group will either be the first nucleoside eluted from the first DEAE-cellulose column, or the last base eluted from the second DEAE-cellulose column.

The nucleotide sequence at the 3′-linked end of bacteriophage MS2 RNA has been elucidated by Glitz (1968) using a similar approach. The end phosphate group was removed by phosphomonoesterase treatment of purified MS2 RNA (molecular weight 1·05 million, i.e. about 3300 nucleotides), and the resulting RNA incubated with ATP (labelled in the γ position with ^{32}P) and polynucleotide kinase. This procedure resulted in the incorporation of 0·2 to 0·8 moles of ^{32}P/mole RNA in the end position (3′-). It was not possible to carry out this labelling without the prior phosphomonoesterase treatment indicating that the 3′-end is naturally phosphorylated. Pancreatic RNase degradation of labelled MS2 RNA was followed by DEAE-cellulose chromatography in 7 M urea, and the fractions assayed for E_{260} and c.p.m. ^{32}P (Figure 3.2). The largest amount of ^{32}P was found in the fifth peak, corresponding to a

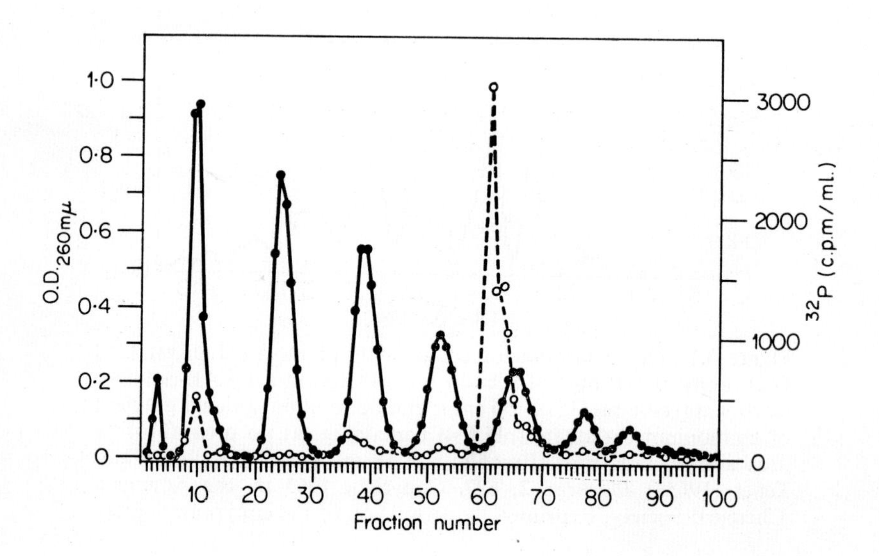

Figure 3.2. Chromatography of polynucleotide kinase labelled RNA after pancreatic ribonuclease degradation. MS2 RNA digest (5 mg.) was taken up on a 15 × 1 cm. diameter column of DEAE-cellulose and eluted with 1300 ml. of 7 M urea-0·02 M tris-HCl (pH 7·5), 0·0–0·4 M linear gradient of NaCl. Fractions (10 ml.) were collected at a flow rate of 0·4 ml./min. (After Glitz (1968) *Biochem.* **7,** 927. Copyright 1968 by the American Chemical Society. Reprinted by permission of the copyright holder)

net charge of — 6 at pH 7·5. This region consisted of a triplet of nucleotides, and later work indicated that the initial sequence could possibly be pppG pPu p py p. The alkaline hydrolysis of this triplet suggested that the sequence was pppG p G p U p.

Alternatives to urea: Sephadex as a matrix and its use for tRNA Fractionation

The difficulties produced by secondary binding forces when using a cellulose matrix can be overcome, as we have seen, by using agents in the eluting buffers which reduce the hydrogen bonding between the bases and the cellulose fibres. This problem can also be approached in another way by using an alternative matrix which does not participate so strongly in secondary reactions. Sephadex, a crosslinked dextran polymer, has therefore been widely used as a support for the ion-exchange groups.

A comparison of the separating properties of DEAE-Sephadex and DEAE-cellulose was carried out by Kawade, Okamoto and Yamomoto (1963) using tRNA molecules with definite amino acid-accepting properties. DEAE-Sephadex was marginally better, but in both systems only valine-tRNA molecules were well separated from the rest. Stephenson and Zamecnik (1962) had already shown that ^{14}C-valine-tRNA could be purified to 90 per cent purity on DEAE-Sephadex, as the valyl-RNA was the first to be eluted from the column (Kawade and coworkers 1963).

Parameters affecting DEAE-cellulose and DEAE-Sephadex chromatography

A detailed examination of the factors which influence the fractionation of yeast tRNA on DEAE-cellulose and DEAE-Sephadex columns was carried out by Cherayil and Bock (1965). They used the low capacity (0·4 meq/g.) DEAE-cellulose and the medium grade capacity (3·3 meq/g.) DEAE-Sephadex A-50, which were washed with either water and M NaCl or 0·1 M HCl, NaOH and M NaCl to remove coloured substances, and then suspended in 0·4 M NaCl buffer. The columns were packed under gravity and equilibrated with buffer containing NaCl and urea, which was pumped through the column at a flow rate of about 10 ml./h. per $cm.^2$ column cross-section. In order to achieve a high resolution between the components of a particular sample during fractionation, the sample was loaded on to the column at that salt and urea concentration which allowed maximum retention (i.e. under conditions where the R_f-values of the various components were close to zero). The fractionated components could then be eluted by slowly increasing the concentration of salt or urea. Each of the two eluting agents could be used independently: a gradient of increasing sodium chloride could be applied in the presence of a constant amount of urea (7·0 M) or, alternatively, the urea concentration could be varied while maintaining the sodium chloride concentration constant (0·34 M). The elution profiles were monitored by $E_{260m\mu}$ measurements against fraction number.

In the absence of urea, tRNA was eluted from DEAE-cellulose between 0·4 and 0·65 M NaCl, and from DEAE-Sephadex between 0·7 and 0·85 M NaCl, with very similar profiles. Magnesium chloride decreases the requirement for NaCl during elution, and if a $MgCl_2$ gradient is superimposed on the NaCl gradient, the resulting elution profile is changed as shown in Figure 3.3. In this example, the tRNA preparation was charged

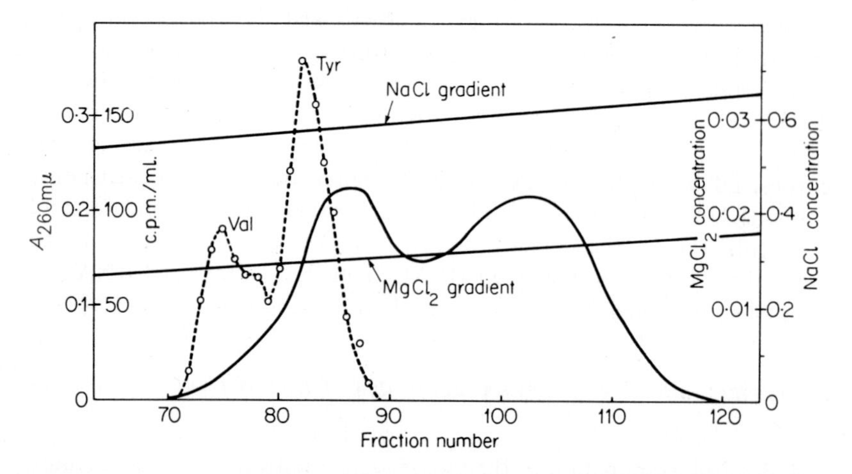

Figure 3.3. Elution of tRNA with NaCl and $MgCl_2$ gradient at pH 5·2. tRNA (14 mg.) prelabelled with [^{14}C] valine and [^{14}C] tyrosine was applied to a DEAE-Sephadex column (1·5 × 100 cm.) equilibrated with 0·5 M NaCl and 0·01 M $MgCl_2$ in 0·02 M acetate buffer. Elution was performed by a gradient linear in NaCl and $MgCl_2$. Fraction size was 6 ml., collected every 30 min. Aliquots of the fractions were dried in planchets and radioactivity was determined in a gas-flow counter. (After Cherayil and Bock (1965) *Biochem.* **4,** 1174. Copyright 1965 by the American Chemical Society. Reprinted by permission of the copyright holder)

with radioactively labelled ^{14}C-valine and ^{14}C-tyrosine prior to fractionation, so that the position of these two tRNA's could easily be detected. The pH was maintained at a low value (pH 5·2) in order to stabilize the amino acid–tRNA complex. The results indicate that all the radioactivity is observed in the early fractions (70–90) of the tRNA profile. Moreover, there are two peaks of radioactivity which Cherayil and Bock (1965) claim are the separate entities of valine- and tyrosine-tRNA molecules. Unfortunately, no further experiments were carried out. However, the effect of urea on the elution of tRNA on a DEAE-cellulose column (1 × 50 cm.) was investigated. After equilibration with 0·25 M NaCl, 7·0 M urea and 0·02 M potassium cacodylate buffer (final pH 7·6), the column was loaded with tRNA (20 mg.) and eluted with a gradient of NaCl (0·28–0·4 M) as shown in Figure 3.4. An aliquot of each fraction was taken and the amino acid-acceptor assay (see Chapter 1) performed for each of the amino acids tryptophan, glycine, valine, tyrosine, lysine, phenylalanine and leucine, of which tryptophan and leucine are shown in the Figure 3.4.

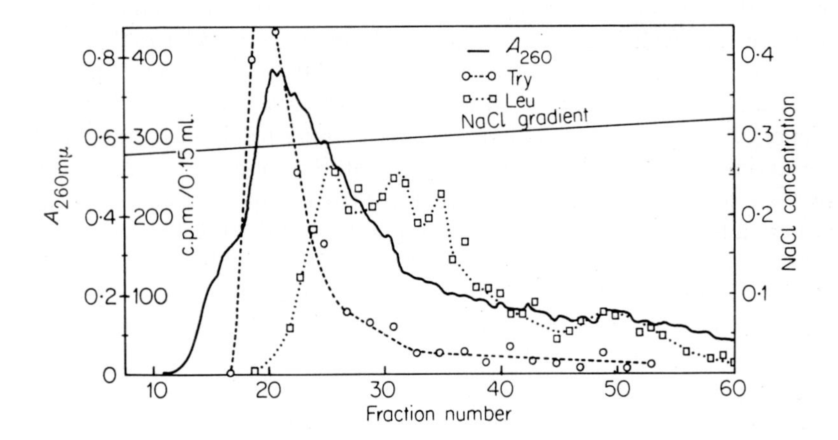

Figure 3.4. Elution of tRNA in the presence of urea. RNA (20 mg.) was applied to a 1 × 50-cm. DEAE-cellulose column equilibrated with 0·25 M NaCl, 7·0 M urea and 0·02 M potassium cacodylate buffer of pH 7·0 (pH of mixture was 7·6). Elution was performed by a gradient of NaCl, 0·28–0·4 M, in a total volume of 500 ml., in 7·0 M urea. Fraction size was 3 ml. collected every 30 min. (After Cherayil and Bock (1965) *Biochem.* **4,** 1174. Copyright 1965 by the American Chemical Society. Reprinted by permission of the copyright holder)

Tryptophan-tRNA eluted first, and the other amino acid tRNA's were subsequently eluted in the order given above. Leucine-tRNA was unusual as it resolved into multiple peaks of a much more diffuse nature, and this elution profile was very sensitive to slight changes in the type of gradient used. If the gradient was sharper, all the tRNA eluted in one peak, whereas if a very shallow gradient was used, the volumes of the eluting medium were inconveniently large.

In the presence of 7·0 M urea, all the tRNA has eluted from the column when the salt molarity reaches 0·34 M. In the absence of urea, all the tRNA remains bound to the column at this salt concentration. It is therefore possible to elute the tRNA from a column with a urea gradient while maintaining a constant salt concentration, and in this case most of the u.v. absorbing material is eluted from a DEAE-cellulose column when a concentration of 5·3 M urea has been reached (pH 7·5). By applying the amino acid-acceptor assay on aliquots from each fraction, it was found that histidine-tRNA was resolved from valine-, arginine- and proline-tRNA's, and that the leucine-tRNA showed three distinct peaks. This type of elution was much less sensitive to gradient conditions, and a much steeper urea gradient could be used without loss of resolution. The early part of the peak, eluted with a urea gradient, contained the arginine- and valine-tRNA's which were almost coincident and poorly resolved. Rechromatography of this region on DEAE-Sephadex at pH 4·5 using a NaCl gradient in 7·0 M urea, however, gave the profile shown in Figure 3.5. The arginine- and valine-tRNA's were resolved, the purity of the arginine-tRNA being about 50 per cent. The arginine- and proline-tRNA's, however, were not resolved under these conditions and, in order to achieve a better separation, rechromatography on DEAE-cellulose using a urea gradient at pH 4·5 was necessary. At this pH the cytidylic and adenylic residues were protonated, which gave an alteration to the interaction of tRNA with the ion-exchanger. Consequently, the components which previously chromatographed as one peak, could be resolved.

Cherayil and Bock (1965) concluded that three principles were involved in the separation of tRNA's from a mixture: (*a*) chain-length of the RNA (when a high urea concentration and dextran matrices were used to exclude non-electrostatic effects, the separation was markedly dependent on the length of the polyanion); (*b*) the amount of protonatable bases, (decreasing the net charge by lowering the pH, resulting in the partial protonation of cytosine and adenine, markedly altered the order of elution of certain tRNA molecules, and indicated differing degrees of protonatable bases); (*c*) degree of interaction with cellulose, (this has already been discussed (Tomlinson and Tener, 1963a), and it is thought that the purine bases interact strongly with cellulose but only weakly with dextran, but however, both types of interaction are overcome by urea).

These three principles can be applied sequentially to achieve the maximum resolution of any one RNA species. For example, fractions obtained using a urea gradient from a DEAE-cellulose column (0·34 M NaCl pH 7·6) can be adjusted to 7·0 M urea, applied to a DEAE-Sephadex column, and eluted with a NaCl gradient in 7·0 M urea (0·35–0·6 M NaCl). Fractions from this second column can then be adjusted to pH 4·5,

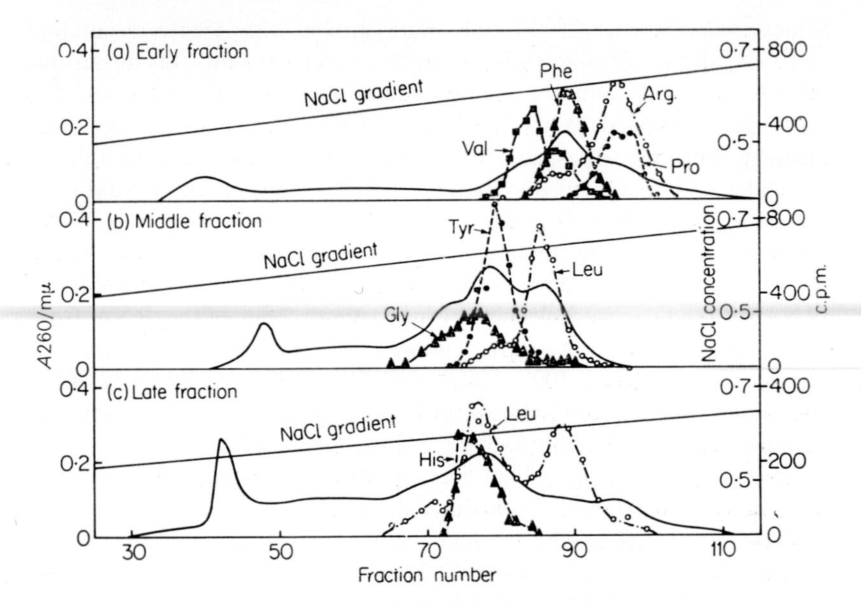

Figure 3.5. Rechromatography on DEAE-Sephadex at pH 4·5. A 1·2 × 100 cm. column was equilibrated with 0·4 M NaCl in 7·0 M urea and 0·03 M acetate buffer, pH 4·5. Fractions a, b and c eluted from DEAE-cellulose column were adjusted to approximately 7·0 M urea with solid urea and to pH 4·5 with acetic acid. The samples were applied to the column and eluted by a linear gradient of NaCl, 0·52–0·70 M in a total volume of 500 ml., in the presence of 7·0 M urea and buffer. Fraction size was approximately 4 ml., collected every 30 min. (After Cherayil and Bock (1965) *Biochem.* **4,** 1174. Copyright 1965 by the American Chemical Society. Reprinted by permission of the copyright holder)

applied to a second DEAE-Sephadex column, and eluted at a constant 7·0 M urea concentration but variable salt (0·52 to 0·7 M NaCl) concentration at pH 4·5. Provided that the effective R_f values of the RNA species are less than 0·05, all the components are absorbed in a narrow zone at the top of the column. Recovery of RNA is almost quantitative, and the acceptor activities estimated have values in the 70–80 per cent range. Aminoacyl-tRNA's can also be separated directly using columns at pH 4·5 with either urea or salt gradients.

Fractionation on DEAE columns relies on the differences between the various amino acid tRNA molecules which pertain under normal conditions. However, alterations in isoleucyl-tRNA, isolated from *E. coli* cells grown in an anaerobic rather than an aerobic environment, can also be detected using DEAE-Sephadex as shown by Kwan, Apirion and

Schlessinger (1968). *E. coli* cells were grown with vigorous aeration (aerobic cultures) or under 95 per cent N_2 and 5 per cent CO_2 (anaerobic cultures) in a medium containing sodium thioglycolate (0·5 g./l.). The cells were harvested in the log phase of growth, and the ribosomes, together with the ribosomal sub-units, removed by centrifugation at 105,000 g. and resuspended in buffer. The supernatant fraction (S-100) was recentrifuged at 105,000 g. for a further 4 h., dialysed against buffer and separated into two fractions by chromatography on Sephadex G-100 (see Chapter 5). The supernatant fraction (S-100) was used to 'charge' tRNA with radioactive isoleucine by incubating 1·2 mg. of the tRNA, from either the aerobic or anaerobic cultures, in 1 ml. of reaction mixture, which also contained 100 μmoles tris-HCl (pH 7·0), 1 μmole ATP, 10 μmoles $MgCl_2$, 10 μmoles KCl, 0·2 mg. bovine serum albumin, 0·24 mg. S-100 and either 0·5 μg. ^{14}C-isoleucine (2 μc/μg.) or 10 μg. ^{3}H-isoleucine (3 μc/μg.). The ^{14}C- or ^{3}H-isoleucyl-tRNA's were then fractionated on a DEAE-Sephadex column (1 × 27 cm.) using a linear gradient, produced by the gradual mixing of two solutions containing 0·5 M NaCl, 0·01 M $MgCl_2$ in 0·02 M potassium acetate buffer (pH 5·1) and 0·75 M NaCl, 0·02 M $MgCl_2$ in 0·02 M potassium acetate buffer (pH 5·1), respectively. Fractions were collected, and their extinction monitored at 260 mμ. Aliquots from each fraction were then treated with cold 5 per cent trichloroacetic acid and the resulting precipitates collected on glass fibre discs. These were washed with cold 5 per cent trichloroacetic acid and 0·1 N HCl, dried, placed in vials containing liquid scintillator and the ^{14}C or ^{3}H estimated.

In a control experiment (Figure 3.6a), aerobic tRNA was separately charged with ^{3}H- and ^{14}C-isoleucine. The two species were then combined and fractionated on a DEAE-Sephadex column as described above. The total tRNA eluted as a broad peak, and the isoleucyl-tRNA was detected as a double peak in the earlier fractions by radioactive counting. The $^{3}H/^{14}C$ ratio was calculated and found to be constant for all fractions, indicating that there were no differences in the amounts of ^{3}H- or ^{14}C-compound eluting at any given molarity.

When aerobically and anaerobically-derived tRNA's were charged with ^{14}C-isoleucine and ^{3}H-isoleucine, respectively, and the two tRNA's combined and fractionated as above, a broad total tRNA peak was again observed (Figure 3.6b), and the aerobic tRNA detected by ^{14}C counts, whilst the anaerobic tRNA was detected by ^{3}H counts. These peaks were closely coincident, eluting at the leading edge of the total tRNA profile described above, but on the descending edge of the ^{3}H curve an additional shoulder was observed. This accounted for 5–10 per cent of the total ^{3}H-isoleucyl-tRNA. The $^{3}H/^{14}C$ ratio was calculated for each fraction and found to be variable. The fractions eluted at low salt concentrations contained less ^3H than ^{14}C, whereas the reverse was true for the fractions eluted at high salt concentration. Kwan, Apirion and Schlessinger (1968) suggested that a modified (rather than a completely new) isoleucyl-tRNA molecule was formed during anaerobic growth.

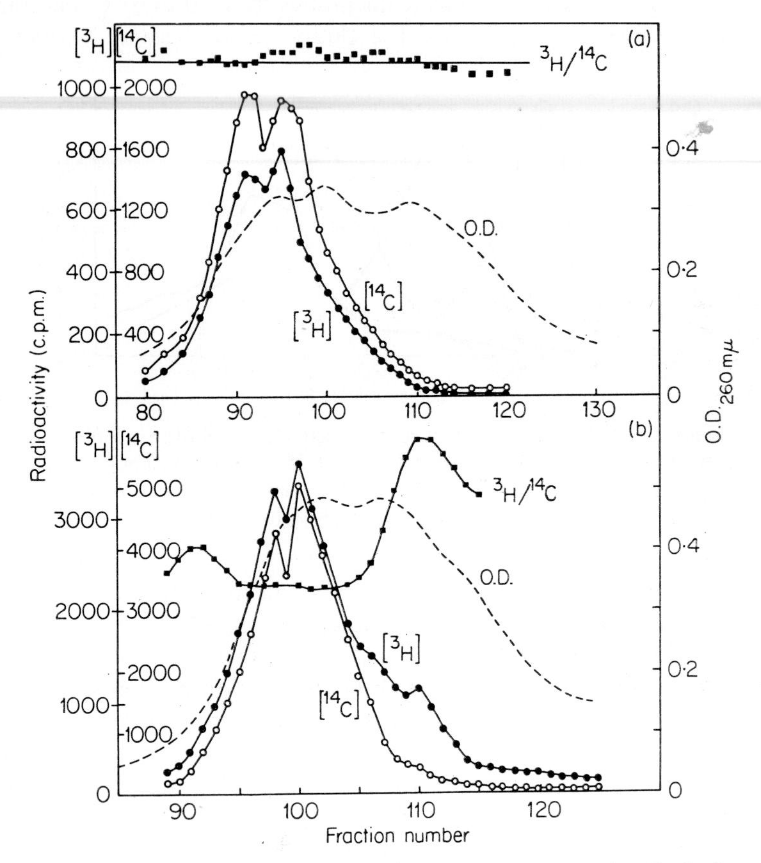

Figure 3.6. Elution pattern of isoleucyl-tRNA on DEAE-Sephadex columns. (a) Aerobic (^{14}C) isoleucyl-tRNA (O.D.$_{260}$ 15·6, 32,000 c.p.m.) and aerobic (^3H) isoleucyl-tRNA (O.D.$_{260}$ 16·0, 12,400 c.p.m.) were mixed and placed on a column. ^{14}C counts per minute (○—○); ^3H counts per minute (●—●); ^3H: ^{14}C (■—■). (b) Aerobic (^{14}C) isoleucyl-tRNA (A_{260} 23·0, 54,900 c.p.m.) and anaerobic (^3H) isoleucyl-tRNA (A_{260} 50·0, 33,500 c.p.m.) were mixed and placed on a column. ^{14}C counts per minute (○—○); ^3H counts per minute (●—●); ^3H: ^{14}C (■—■). (After Kwan, Apirion and Schlessinger (1968) *Biochem.* **7**, 427. Copyright 1968 by the American Chemical Society. Reprinted by permission of the copyright holder)

Effect of temperature of fractionation

Baguley, Bergquist and Ralph (1965) fractionated tRNA from brewers' yeast using a column of DEAE-cellulose (1·5 × 40 cm.), and linear gradients of NaCl (0·6–1·0 M) in buffer containing EDTA (2mM) and sodium acetate (20mM). The columns were surrounded by a water jacket to allow fractionation at various temperatures from 20 to 65°C, and their results are shown in Figure 3.7. The elution profile became broader as

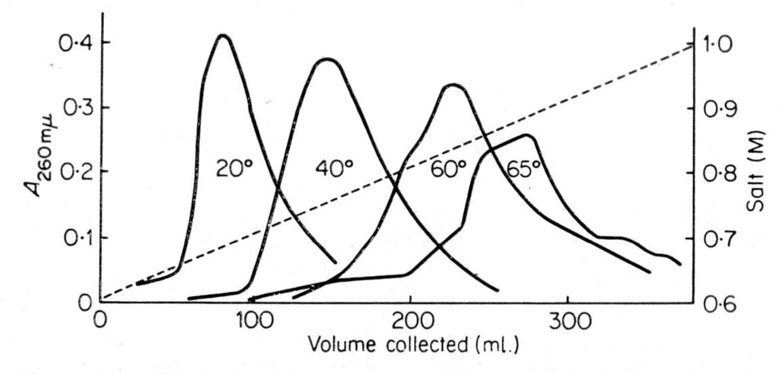

Figure 3.7. Effect of temperature on yeast tRNA elution profile and position from DEAE-cellulose columns (1·75 × 18 cm.) tRNA (1 mg.) was eluted with a linear NaCl gradient at the temperatures shown. (After Baguley, Bergquist and Ralph (1965))

the temperature was increased, and greater salt molarities were required to elute the RNA. The amino acid-accepting properties of the various fractions were also tested, and it was found that, by the appropriate choice of temperature and salt gradient, the elution position of specific tRNA's could be controlled so that certain types were retarded more than others. Also, if a ten-fold enriched lysine-tRNA, which was contaminated with valine- and isoleucine-tRNA's, was fractionated on heated DEAE-cellulose, then pure lysyl-tRNA (81 per cent purity) could be resolved. Subsequently Bergquist, Baguley, Robertson and Ralph (1965) plotted the position of about thirty-five tRNA species many of which were multiple tRNA's for a single amino acid. They concluded that the fractionation on heated DEAE-cellulose columns depended on the degree of unfolding that occurred in the various tRNA species at different temperatures, the more unfolded tRNA's forming a greater number of bonds with the column matrix. Fractionation also depended on the GC content of the tRNA, a higher GC content causing increased base-pairing. In this respect, alanine-acceptor tRNA eluted first from the columns, and is also reported to have a high GC content.

BND-cellulose and BD-cellulose column chromatography

Secondary, non-ionic, interactions on DEAE-cellulose columns can be increased by modifying the DEAE-cellulose (Gillam and coworkers, 1967). Benzoyl and naphthoyl groups were introduced on to the DEAE to give a type of exchanger with an affinity for lipoidal groups. The cellulose produced, benzoylated DEAE-cellulose (BD-cellulose) or benzoylated naphthoylated DEAE-cellulose (BND), bound polynucleotides more strongly, despite the fact that substitution of the ionic groups on the cellulose for non-ionic aromatic groups should have decreased the capacity for ion-exchange. As with DEAE-cellulose columns, the secondary forces present on BD-cellulose could also be eliminated by urea or alcohols, as these substances disrupt both hydrogen and hydrophobic bonds.

Gillam and coworkers (1967) fractionated tRNA (5g.) from yeast on a BND-cellulose column (4.3 × 110 cm.) eluted with a 10 l. linear gradient of sodium chloride (0·45–1·0 M) in 0·01 M magnesium chloride (Figure 3.8). Several u.v. absorbing peaks were noted, and two peaks of histidine-tRNA activity could be detected when assayed with ^{14}C-histidine. Several regions from these peaks, when pooled and rechromatographed, eluted at the expected salt concentration, which showed the fractionation was reproducible.

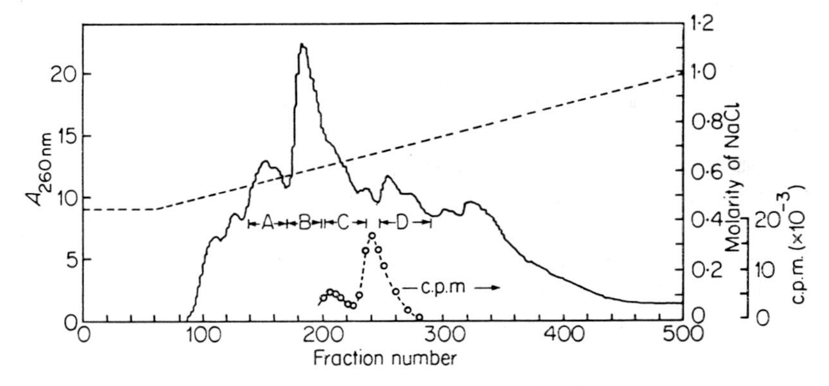

Figure 3.8. Elution profile from a column (4·3 × 110 cm.) of BND-cellulose loaded with 5 g. (72,000 A_{260} units) of tRNA from yeast (Calbiochem). The sample was applied in 200 ml. of 0·45 M sodium chloride–0·01 M magnesium chloride, and eluted with 10 l. of sodium chloride (a gradient from 0·45 to 1·0 M)–0·01 M magnesium chloride. Fractions were 20 ml./10 min. Absorption of each fraction at 260 nm. (solid line); acceptor activity for (^{14}C)-histidine (dotted line). (After Gillam and coworkers (1967) *Biochem.* **6**, 3043. Copyright 1967 by the American Chemical Society, Reprinted by permission of the copyright holder)

Aminoacylated tRNA's were fractionated on BND-cellulose using a pH 5 buffer containing sodium acetate (0·05 M). Figure 3.9 shows the fractionation of enzymatically aminoacylated bakers' yeast tRNA

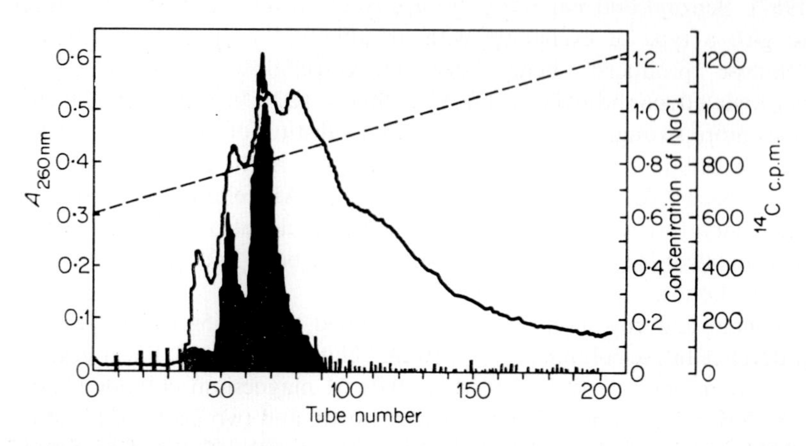

Figure 3.9. Chromatography of (^{14}C) alanyl-tRNA (19,000 c.p.m.) prepared by the standard esterification procedure from tRNA of bakers' yeast (240 A_{260} units) on a column (1 × 110 cm.) of BND-cellulose. Elution was with a gradient of sodium chloride (a total of 1 l.) from 0·6 to 1·2 M, containing 0·01 M magnesium chloride and 0·05 M sodium acetate (pH 5·0). Fractions were 5 ml./10 min. Solid line: absorbance at 260 nm.; blackened area: counts per minute. (After Gillam and coworkers (1967) *Biochem.* **6,** 3043. Copyright 1967 by the American Chemical Society. Reprinted by permission of the copyright holder)

(^{14}C-alanine) in acetate buffer superimposed by a 0·6–1·2 M sodium chloride gradient. The pattern of elution was not substantially different from that observed previously, and two peaks of radioactivity were found early in the profile, corresponding to two separable species of alanyl-tRNA. Random hydrolysis of the alanyl-tRNA could not be completely prevented even at pH 5·0, so that the background radioactivity, due to free ^{14}C-alanine was high prior to the main peak, but decreased to low values subsequently.

The chromatographic separation on BD-cellulose of the amino acid tRNA's on a preparative scale has been extensively investigated by Gillam and coworkers (1967). Eighteen of the common amino acids were used in assays on the fractions obtained after chromatography of 5 g. of tRNA from brewers' yeast on a BD-cellulose column (3·2 × 110 cm.), eluted with a linear NaCl gradient (0·45 to 1·0 M) in 0·01 M $MgSO_4$. The elution position of each amino acid-acceptor activity was plotted, and in several cases multiple species for one specific amino acid were found. For example, glycine gave two major and two minor peaks of activity, the two major peaks eluting very early in the absorbance profile. Rechromatography of the first of these regions (at low pH) did not resolve the glycine activity from that of isoleucine. However, the second major peak, when rechromatographed at pH 3·5 in the presence of 0·005 M EDTA, separated into two peaks, one coincident with alanine activity, and the other reasonably well resolved from the first. Phenylalanine activity gave a unique elution profile in that no detectable activity could be found in the salt-gradient elution, and treatment of the column with 10 per cent (v/v) 2-methoxyethanol in 1·0 M NaCl was required before it could be released. This fraction contained very little other acceptor activity and was enriched twelve- to fourteen-fold.

The effect of tRNA conformation on binding

Open-stranded chains bind more strongly to partially benzoylated DEAE-cellulose and BND-cellulose than do the more compact tRNA molecules in the native state (Baguley, Bergquist and Ralph, 1965). This is because the reactive groups on the exchangers are distributed at random along the cellulose fibre (so as not to break down the forces holding the fibre together), and a single-stranded open polynucleotide chain has a greater chance of interacting with a larger number of these sites than does the compact molecule. If the secondary structure of a tRNA molecule is therefore reduced (for example by an increase in temperature or removal of Mg^{2+} ions), the intensity of tRNA binding will increase and the molecule will only be eluted by a high salt concentration. Ribosomal RNA also cannot be eluted from partially benzoylated DEAE-cellulose by the usual salt gradient, and even increasing concentrations of 2-methoxyethanol fail to release more than 30 per cent of the rRNA. This is again due to the open structure of rRNA.

Introduction of artificial aromatic groups on to tRNA to facilitate separation

The degree of benzoylation and naphthoylation of the DEAE-cellulose is fairly critical. Purely naphthoylated DEAE-cellulose has too great an affinity for RNA, and no material can be eluted from these columns by salt gradients. Thus in BND-cellulose, the naphthoyl groups are kept to an estimated 9 per cent (on a molar basis) of the content of benzoyl groups.

Aminoacyl-tRNA's are usually chromatographed on BD-cellulose using buffers maintained at pH 4. However, tyrosyl-, tryptophanyl- and phenylalanyl-tRNA's can only be eluted by ethanolic solutions. This is because the aromatic amino acid on the tRNA molecule binds with an increased affinity. This aids the separation and purification of these molecules from those which do not possess such groups (Maxwell, Wimmer and Tener, 1968). The method can also be applied generally for isolating specific tRNA's if aromatic groups are incorporated artificially onto the reactive amino group of the aminoacyl-tRNA under investigation, since the substituted compound can be separated from the remaining species by passage through BD-cellulose (Gillam and coworkers, 1968). A mixture of tRNA's is first passed through BD-cellulose (eluted with M NaCl) to separate material which already has an affinity for the column, and is then charged with the specific amino acid of interest, using an aminoacyl-tRNA synthetase. There are several specific points worth noting at this stage. Firstly, the objective is to purify a tRNA and the purity of the amino acid used for esterification is therefore crucial. Secondly, the synthetase preparation must be free both of amino acids and amino acid biosynthetic ability. Once the amino acid is linked to the tRNA, the amino group of the ester is substituted with an aromatic acyl group, and phenoxyacetyl and naphthoxyacetyl are commonly used (Gillam and coworkers, 1968). In all the experiments described the esters of *N*-hydroxysuccinimide were used as acylating agents, and the active esters were prepared very simply by shaking the two reactants together in a stoppered flask, with dry dioxan as the solvent. The two esters have a low solubility in water but this does not prevent the rapid substitution of the aminoacyl-tRNA. Very little reaction occurs with uncharged tRNA as shown by the following control experiment. RNA pellets were suspended in 0·1 M triethanolamine hydrochloride and 0·01 M magnesium chloride (pH 4·3), and treated with the active ester which had been dissolved in dry tetrahydrofuran. The pH was quickly adjusted to 8·0 by N sodium hydroxide, and after 10 min. to pH 4·5 with glacial acetic acid. The presence of water and triethanolamine seems to limit the attack upon the random hydroxyl groups on the RNA molecules. The RNA molecule under investigation should now contain an aromatic group not possessed by the other RNA species. In order to ensure that random acylation has not occurred, a second acylation can be carried out on a sample of RNA which has not been substituted with an aminoacyl derivative. Any random acylation would then be revealed in a comparison of the results from the two samples. Alternatively, the sample can be taken through the acylation procedure before charging with the amino acid, and the products subjected to BD-cellulose chromatography to remove spurious acylated derivatives. Only the 'pure' material is used for the subsequent stages. The molecules containing the aromatic groups are subsequently separated from those without such groups on BD-cellulose, since the former are more strongly adsorbed. The sample is loaded and eluted in the usual way, and it should be found that little or no activity remains in the major profile for the tRNA under investigation, but all the activity is eluted on treating the column with M NaCl, 0·01 M $MgCl_2$ in 9·5 per cent (v/v) ethanol. In order to make identification easier, a ^{14}C label was included in the amino acid (aspartic acid) under investigation, into which a phenoxyacetyl group was introduced (Figure 3.10a). The rechromatography on BD-cellulose of this ethanolic fraction (following the removal of the esterified amino acid by incubation in tris buffer at pH 8 or 9) is also shown. This method gives preparations which are reasonably pure for single tRNA molecular species. However, as shown in Figure 3.10b some contamination, in this case by tyrosyl-tRNA, has resulted from the various reactions (Gillam and coworkers, 1968). Also, in some cases, one amino acid will esterify several specific tRNA's, and hence this procedure will, in the absence of additional specific synthetases, result in the isolation of a genus of tRNA's specific for that amino acid. However, it should be realized that it is often possible to separate these various components by an additional chromatographic step, or by employing some of the other separation methods described.

Sedat, Kelly and Sinsheimer (1967) fractionated a nucleic acid mixture extracted from *E. coli* infected with bacteriophage MS2, using BND-cellulose, and found that the secondary structure of the nucleic acids determined the degree of binding to the aromatic groups bound to the cellulose matrix. DNA was separated from 16S and 23S RNA but tRNA was not separated from DNA. More recent studies using BD-cellulose (Sedat, Lyon and Sinsheimer, 1969) have allowed the purification of a rapidly-labelled mRNA fraction from *E. coli* approximately fifteen-fold. The conditions of elution (presence of organic compounds, e.g. dimethyl sulphoxide, dodecyl sulphate) are more complex than that used for fractionation of tRNA molecules, and the physical basis for the separation between the various rRNA molecules, mRNA and DNA is not yet fully understood.

Summary

Although anion-exchange resins have been used to fractionate ribosomal RNA precursors, DNA–RNA hybrids and more recently DNA and mRNA, their most important use has been with oligonucleotides and aminoacylated-tRNA molecules. The fractionation of tRNA's on DEAE

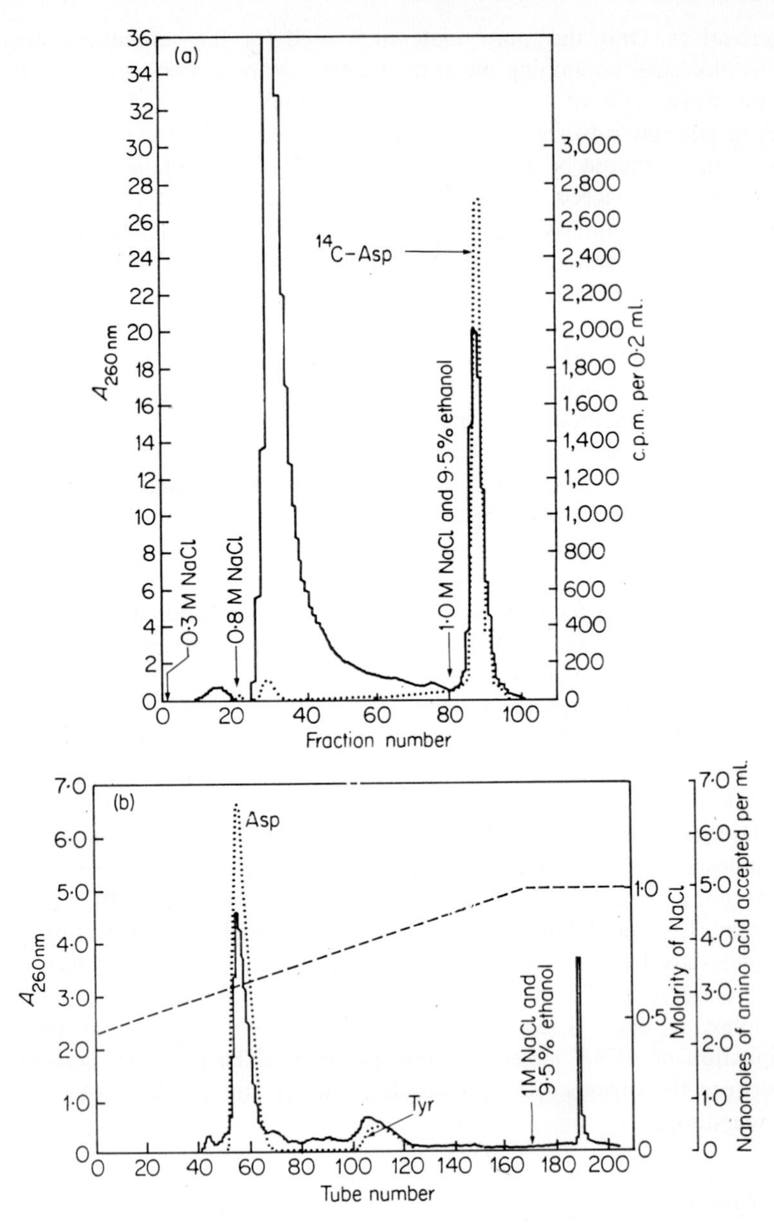

Figure 3.10. Preparation of tRNA^{Asp} by the phenoxyacetylation procedure. (a) Aspartyl-tRNA prepared using purified synthetase with 1 g. of tRNA from which the EF and sham-acylated fractions had been removed was phenoxyacetylated after addition of (^{14}C) aspartyl-tRNA prepared similarly from 5 mg. of tRNA. The tRNA was applied to a

columns not only depends on the ion-exchange properties (since all the various tRNA's are similar in size and charge) but to a large extent on the weaker secondary forces of interaction which depend on the base composition of the polynucleotide. In this respect, purine bases bind more strongly than pyrimidine bases.

Careful control of these secondary forces and choice of pH and salt gradient provide an effective means of fractionating many tRNA's. The secondary forces may be reduced by changing from a cellulose to a Sephadex matrix, or by including a hydrogen-bond breaking agent such as urea in the eluting buffer. Alternatively the forces can be increased and used as the main basis of fractionation by increasing the temperature. This results in the uncoiling of the base-paired regions of the tRNA's, which leads to an increased binding to the matrix. Since the degree of base-pairing is greater for those molecules having a high GC content, differences in base-composition between the various tRNA's will give rise to different degrees of single-strandedness and hence different binding intensities. Additional interaction forces can be introduced between the polynucleotide and cellulose matrix if the pH of the eluting buffer is approximately 4·5, the adenylic and cytidylic acids being protonated at this pH.

The forces between tRNA molecules and BD- or BND-cellulose matrices are stronger than those in the case of non-aromatically substituted matrices due to the presence of hydrophobic bonds in addition to hydrogen bonds. The strength of the hydrophobic bonds, however, can be decreased by urea, or increased and utilized for fractionation by uncoiling the tRNA by heat or decreasing the Mg^{2+} concentration.

Since hydrophobic bonding is important in the case of BD- and BND-cellulose, the tRNA molecules accepting the aromatic amino acids, tyrosine, phenylalanine and tryptophan can be preferentially separated

---

column of BD-cellulose in a solution A and subsequently eluted with solutions B and E as indicated by arrows. Solid line: A_{260}; dotted line: radioactivity. (b) The peak eluted by solution E above was stripped of esterified amino acid and chromatographed on a column (1·4 × 110 cm.) of BD-cellulose. Elution was with the indicated (dashed line) gradient of concentration of sodium chloride (total of 3 l.), 0·01 M in magnesium chloride. Flow rate was 1·7 ml./min. At fraction 172 elution with 1·0 M sodium chloride–0·01 M magnesium chloride in 9·5 per cent (v/v) ethanol was started. The volumes of fractions decreased after this point. Solid line: A_{260}; dotted line: acceptor activity for aspartic acid; labelled dashed line: acceptor activity for tyrosine. (After Gillam and coworkers (1968) *Biochem.* **7**, 3459. Copyright 1968 by the American Chemical Society. Reprinted by permission of the copyright holder) from other tRNA's by virtue of their greater binding capacity. This property has been applied further by artificially introducing an aromatic group on to the amino acid attached to a particular tRNA. This allows the separation and isolation of pure tRNA molecules which can be utilized for example in *in vitro* protein synthesizing systems and for sequence analysis.

CHAPTER 4

Partition chromatography and countercurrent distribution

General introduction

The separation of the different components within a mixture by differential partition between two immiscible liquids, has long been an important and fundamental technique in chemistry and biochemistry. The method depends on the fact that solutes can have different solubilities in these two liquids so that, after sufficient equilibration, the amount of solute in one is greater than in the other. In this way, a 'concentrating effect' is obtained and, because the two liquids are immiscible, they are easily separated. Simple, single-stage, concentrations of this type have their uses, and examples will be given later. They are extremely useful as only very simple apparatus is required, usually a single separating funnel. However, to effect such a separation, the difference in solubilities must be great, and other components of the original mixture must not be transferred to any significant extent. This rarely happens with biological molecules. Quite often a mixture contains a range of molecules with very similar solubility properties, and it is very difficult to separate one particular species of interest.

The obvious answer, when the solubility difference between two solutes is small, is to perform the liquid–liquid equilibration and separation several times, keeping one of the phases constant, and using repeated fresh quantities of the other phase. Eventually a very small amount of the component with the lowest solubility will remain in a purified form, but the yield (if the solubilities are close) will be very small. If, however, after the first extraction, the upper phase is not only replaced by fresh solvent in the first tube, but also transferred to fresh lower phase in a second tube, a second equilibration will result in the components moving back into the lower phase with the same distribution as existed originally. In Figure

4.1 the original tube 1 contains two phases A and B with two solutes ○ and ●, the open-circle molecules being more soluble in A than in B, while the filled-circle molecules are more soluble in B. When phase B

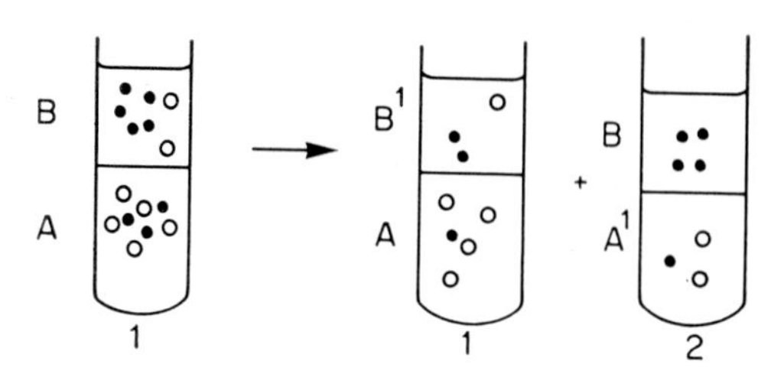

Figure 4.1. Separation of two types of molecule using their different solubility properties in solvents A and B (see text)

(after equilibration) is moved on to fresh phase A (A^1), the components in B equilibrate so that the open-circle molecules move into A^1 and the filled-circle molecules remain in B. Fresh phase B is simultaneously added to tube 1 (B^1) and reequilibrated. This results in a few open-circle molecules, but a greater number of filled-circle molecules moving into B^1. In total phase A ($A^1 + A$), there are six open-circle molecules and only two filled-circle molecules, while in phase B ($B^1 + B$) there is only one open-circle molecule but six filled-circle molecules. A considerable separation has therefore been achieved. If this process were continued for a few more tubes, then eventually all the filled-circle molecules would be found in phase B, while all the open-circle molecules would be found in phase A, a situation which was not obtained in tube 1. This type of procedure is known as countercurrent distribution (CCD), and has proved extremely useful in several types of biochemical separations. The type of solvent systems vary widely according to the nature of the solute molecules, but certain fundamental theoretical concepts apply in all cases.

Theory of countercurrent distribution

The CCD system is very closely related to the 'theoretical-plate' concept of chromatography, except that in this case the 'plates' are real, and have physical existence in the form of the unit cells of a CCD machine. The equilibration of any given solute between the two phases is conveniently carried out by shaking or agitating, so that one solvent becomes dispersed within the other, thus increasing the surface area between them. This allows a more rapid distribution of the solute than would be otherwise achieved. When the equilibration process is considered complete, the agitation is stopped, and the two liquids allowed to separate. One phase is then transferred quantitatively to the next cell in the machine, where it is mixed with a fresh volume of the stationary phase. For convenience, it is usually the upper phase which is moved, whereas the lower phase remains. This gives rise to the often quoted 'mobile phase' and 'stationary phase' definitions. New upper or mobile phase is applied to the first cell, and the whole process repeated until the first portion of upper phase reaches the last cell of the machine. The partition coefficient (K) for countercurrent systems is usually defined as

$$K = \frac{\text{Concentration of solute in upper phase } (C_u)}{\text{Concentration of solute in lower phase } (C_l)}$$

If M_u is the total amount of solute in the upper phase and M_l is the total amount of solute in the lower phase, then the ratio of these two values gives the distribution ratio (G): $G = M_u/M_l$. This is of course related to the ratio of the volumes of liquid (R) in the upper (V_u) and lower (V_l) phases so that if $R = V_u/V_l$ then $G = K \times R$.

In countercurrent systems, R is known and can be varied so as to alter the movement of a given component through the cells without changing K.

The width of any zone moving through the system can be calculated from the properties of the binomial distribution, from which the standard deviation (δ) of the distribution is given by the equation $\delta = \sqrt{nG/G+1}$ where n is the number of cells. However, since 99·9 per cent of the solute will be found within a region extending for 3 δ on either side of the maximum value, it is considered practical for all normal purposes to take the width (W) of the zone as $W = 6 \times \sqrt{nG/G+1}$. This equation is only valid when the zones have a symmetrical profile, i.e. for all values of n where $G = 1$, but is not valid for marked asymmetry and for small values of n, when G departs considerably from unity.

The width of a given zone is therefore proportional to $\sqrt{n}$, whereas the distance between any two zones is proportional to n. Thus by increasing the number of transfers, the resolution of peaks is also improved, and the number of cells a zone occupies as a percentage of the total number of cells becomes smaller, thus: when $G = 1$, for $n = 50$, $W = 21$ and the percentage is 42. However, when $n = 100$, $W = 30$ the percentage is about 30, and when $n = 1{,}000$, $W = 95$, the percentage drops to 9·5. It can therefore be seen that, even under ideal conditions, when $n = 100$ only three components could be separated with 99·9 per cent purity, whereas at $n = 1{,}000$ the maximum number of separable components would be ten.

In one complete CCD process, the best resolution of two components x and y will be obtained when their distribution ratios satisfy the relation: $G_x \times G_y = 1$, and since K_x and K_y are fixed by the solvent system employed only variations in the volume ratio shown earlier can bring about this type of adjustment. The volume ratio required (R_0) is given by the equation: $R_0 = 1/\sqrt{(K_x \times K_y)}$. In most equipment R_0 can be varied between 0·5 and 2·0.

Practical considerations

Countercurrent distribution apparatus usually appears very complicated and, although not very expensive by today's standards, is not cheap and requires a reasonable amount of maintenance. The basic cellular units are usually made of glass, which can be clipped together for convenience on the shaking and transfer assembly. The original equipment was either of the drum or U-tube type, but these have now been superseded by the decantation system in which, after agitation and settling, the upper phase is decanted into the next cell and replenished in a similar manner. A simple system for the transfer of the upper phase only is shown in Figure 4.2. The two solvents containing the solutes are equilibrated in tube X, while the apparatus is rocked back and forth in position A. The apparatus is stopped and tilted slightly to the right, and the two phases allowed to separate. When this separation is complete, the unit is fully rotated into position B, and the upper phase decants along tube Y into the containing vessel Z. On rotation back to position A again this phase moves along tube Z, down tube V and into tube W. Thus the net effect is to move the upper phase from unit (1) into unit (2). Individual units may be sampled through the stoppered tube at the end. The volume in the lower phase is fixed, but the volume in the upper phase may be varied between certain limits, and in most equipment a phase ratio of 2·5 : 1 can be obtained.

In a number of the more modern types of apparatus, the decantation units are designed so that both the upper and lower phases can be transferred in opposite directions. One such system is shown in Figure 4.3. In this equipment agitation takes place in the chamber X with the unit in the horizontal position as shown. After phase separation, the unit is rotated through 90° in the direction of the arrow, so that the upper phase moves through tube Z into the decantation chamber Y. A subsequent rotation to a position 200° from the original horizontal position in the direction of the arrow, transfers the lower phase from chamber X into the preceding chamber Y (i.e. from X_1 into Y) through the tube W. If the apparatus now continues through to a complete 360° turn, then both phases are moved from Y into succeeding chambers X (i.e. from Y into

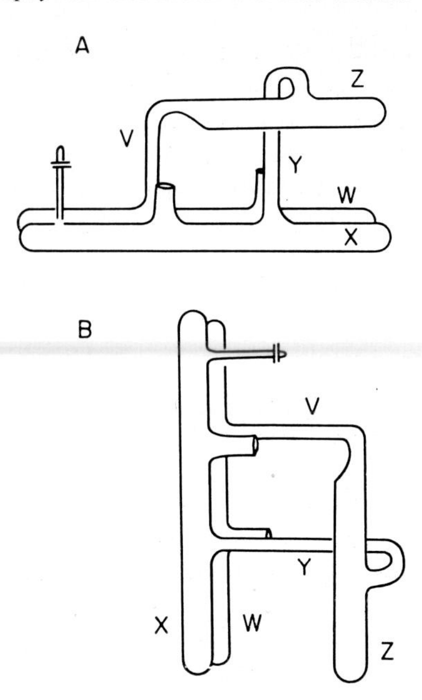

Figure 4.2. A simple apparatus for the transfer of the upper phase in a countercurrent distribution system (see text)

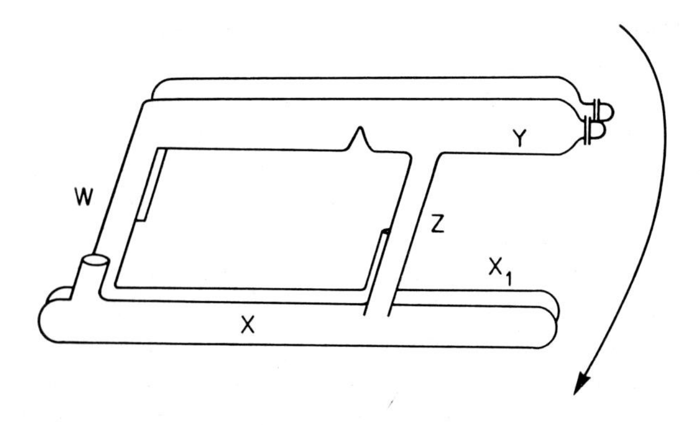

Figure 4.3. A decantation unit for the upper and lower phases in a countercurrent distribution system (see text)

X_1). Summarizing, a full turn of the unit returns the lower phase to the chamber X (from where it came), while the upper phase has been transferred to the succeeding chamber X (i.e. from X to X_1).

The lower-phase transfer can be effected by rotation through 200° in the direction of the arrow from the original position, as in the first case, when the Y chambers contain the upper phase from the preceding X chambers, and the lower phase from the succeeding X chambers. The cycle can then be completed by a 200° rotation back in the opposite direction to the original position. Both phases flow through tube Z into X. The upper phase has remained in its original position, while the lower phase has been transferred from the preceding X chamber (i.e. X_1 to X). The little indentation in the chamber Y prevents the phases flowing in the wrong direction.

This, and other types of apparatus which transfer both phases in opposite directions simultaneously, are usually fully automatic. One of the ways of obtaining maximum separation of a given component from a mixture is to feed the solute into the centre of the distribution bank, and have the system arranged so that the required component remains stationary in the centre, while the two phases carry those substances which are not required, in the two phases moving towards opposite ends of the equipment. Any 'drift' of the component away from the centre can be corrected by arranging more transfers of one phase than of the other, e.g. three moves of the upper phase, with a corresponding two moves of the lower phase, instead of three each.

Solvent systems

This is the critical part of the whole technique, and considerable attention must be given to the choice of solvents in order to obtain maximum separation. Also, because of the time and effort involved in operating the CCD system, it is inadvisable to commit samples to CCD fractionation until the solvent system has been fully tested. Solvent systems for CCD work must separate quickly into two bulk liquids from a finely divided suspension of one phase within the other, and this property must be retained in the presence of quite large quantities of the solutes. In order to achieve this, the phases must differ in density, and neither should have a high viscosity. However, the differences should not be such that the interfacial tension promotes the formation of stable emulsions. The presence of fatty acids or other lipids for instance may lead to stable emulsions which are difficult to separate. Emulsification is often encountered early in the separation, when the local concentration of these substances is high, and may be partially overcome by distributing the solute equally between several tubes at the beginning of a run. If this is less than a 5 per cent distribution of the total number of cells, then the final pattern is not materially affected. Another way to overcome the problems in emulsification, is to carry out the first few transfers manually, outside the main system, and break the emulsions by centrifugation. Once the concentration dependence of this emulsification has been sufficiently reduced, the solute-containing tubes may be transferred back to the machine. The formation of emulsions may also be prevented by eliminating certain impurities, such as solid particles which frequently enhance emulsion formation, and dialysis is also useful in some cases.

The addition of small amounts of specific substances, such as higher alcohols or certain silicones, also may prevent emulsion formation. These substances will of course behave as do other solutes, and distribute during the course of the separation. It is interesting that the problem caused by emulsions is not often mentioned in the literature and in practice they can seriously limit the applications of CCD to natural systems. In many cases, the separation of the two phases is excessively long, and there is evidence that the attainment of equilibrium between phases is considerably decreased by emulsion formation. This gives rise to non-ideal distributions, with the resulting skewed partition curves frequently observed in emulsifying systems. Rapid separation of aqueous and organic solvents can often be aided by either increasing the density of the lower aqueous phase with salts, or by using a high ratio of organic to aqueous phase.

Countercurrent distribution of nucleic acids

Two major systems utilize the partition properties of nucleic acids in immiscible solvents as a CCD fractionation technique; aqueous/organic and aqueous/polymer two-phase systems. In the former, the nucleic acids distribute between an upper aqueous phase, either as the salt of an alkali metal (Apgar, Holley and Merrill, 1962) or as a tributylamine salt (Zachau and coworkers, 1961), and an organic phase, whereas in the second, the solvent for both phases is water, but the immiscibility is produced by the presence of high molecular weight polymers such as dextran (Albertsson, 1965a).

Countercurrent distribution techniques were probably first applied to nucleic acid fractionation by Holley and coworkers (1961), who intensively studied the separation, purification and structural determination of tRNA molecules. The ideas concerning tRNA and its subfractionation (i.e. to species specific tRNA for each amino acid), have already been discussed. Transfer RNA provides a fruitful area for research for several reasons. The molecules are easy to identify as they will specifically take up amino acids which, if radioactively labelled, can be detected in small quantities. Their function in protein synthesis *in vitro* therefore can be easily studied. They are among the smallest of all nucleic acid species in general being only about 75 nucleotides in length, and, potentially, therefore easier to use for sequence analysis. Finally, although containing unusual bases (which might even aid sequence analysis), they contain no protein and therefore have a biological conformation dependent solely on the nucleotide sequence. It is not surprising, therefore, that these molecules have received much attention in the literature, and a great deal is now known about them.

A simple experiment to illustrate CCD

Holley and coworkers (1961) described a simple experiment for the fractionation of tyrosine- and valine-acceptor transfer RNA's and this will serve to illustrate certain fundamental experimental points in CCD work. They used the transfer RNA prepared by the phenol method of Kirby (1956) from fresh bakers' yeast, having separated it from the bulk of higher molecular weight RNA by DEAE-cellulose chromatography. The separation of tyrosine- and valine-acceptor fractions was then accomplished by six transfers in a countercurrent distribution system. For such a small number of transfers it was, of course, not necessary to use a complex CCD machine, as all the processes could be carried out manually in centrifuge tubes. The two phase solvent system consisted of a solution of 22·2g. K_2HPO_4 and 34·8g. NaH_2PO_4 in 0·001 M $MgCl_2$ to give a total volume of 200 ml., which was mixed with 20 ml. of formamide and 80ml. isopropyl alcohol at 25°C. Yeast tRNA gave a gross partition coefficient close to unity in this system. The upper and lower phases (10ml. each) were then mixed with 50mg. of the tRNA at 25°C, and shaken intermittently for at least 15 min., to allow the RNA to dissolve. The contents of the mixing vessel were subsequently transferred to a centrifuge tube (preferably glass), and spun briefly to separate the layers. It will be remembered that the partition coefficient, $K = C_u/C_l$, is dependent on the concentration of the solute in the upper and lower phases, which is in turn dependent on the volume ratio, i.e. $G = K \times R$, and that it is the volume ratio which can be varied to ensure the required degree of separation. In the present case, this can be checked and corrected before proceeding with the CCD fractionation, and provides a useful check for the first stages of many more complex fractionations. Since the partition coefficient varies with the temperature of the solutions and slight changes in the ionic composition of the solvents, the actual partition coefficient of each experiment can be measured by withdrawing portions of each of the two phases after the first distribution, and measuring the extinction at 260 mμ (in the example cited the aliquot has to be diluted 0·1 to 0·3 ml. to give a measurable value). These readings are converted to concentrations of RNA or, as they are proportional to the concentrations in the extinction range measured, simply divided by each other. If the value of K, determined in this way, differs from the required value of 1·0, it is possible to readjust the volumes of the two layers so that about half the RNA is present in each. For example, if the observed K is 1·1, then the volume in the lower layer must be increased by 1ml., i.e. 10 to 11 ml. to correct for this, and this quantity used in the lower layer for the subsequent experimentation. This is the first of the fundamental principles illustrated in this experiment.

The CCD experiment now proceeds as follows: six tubes (40ml. glass centrifuge tubes) are numbered 1 to 6 and the quantity of lower layer (determined as above) pipetted into each. The upper layer of tube 0 is transferred to tube 1, and 10ml. fresh upper layer added to tube 0. The two tubes are then shaken and centrifuged. The upper layer of tube 1 is transferred to tube 2, the upper layer of tube 0 is transferred to tube 1 and a fresh 10ml. of upper layer added to tube 0. The contents of the three tubes 0, 1 and 2 are again shaken and centrifuged, and the upper layers each moved to the next tube. This demonstrates the second principle, that of unidirectional flow of one layer. When the original upper layer of tube 0 reaches tube 6, all the seven tubes have two layers and the procedure is complete. The contents of each tube are then transferred to separate dialysis sacs, and dialysed twice against deionized water for 1h., 3h. against 0·0003 M $MgCl_2$ solution, 15h. against deionized water and 2h. against glass-distilled water. This procedure results in a final volume of about 5·0 ml. for each dialysis sac, so appropriate precautions should have been taken when preparing the sacs to allow for expansion. The volume of each fraction is now reduced ten-fold by rotary evaporation (water bath 45°C), and the reduced volume transferred, with the washings of the flask, to a small test tube, and the final volume adjusted to 5ml. An aliquot of each fraction is taken for assay and its extinction measured at 260 mμ.

One type of assay for amino acid-acceptor activity has already been described (Chapter 1), However, in his original paper, Holley quoted another method in which all the reactions were carried out in test tubes, and the activated RNA's precipitated on to aluminium planchets for radioactive counting. All the tubes contained RNA in approximately equal amounts, as measured by absorbance at 260mμ, but the amino acid-acceptor assays showed significant differences (Holley and coworkers, 1961). The valine-acceptor activity had a maximum in tube 1, and the bulk of the activity was in tubes 0–3. In contrast, tyrosine activity had a maximum in tube 5 and no activity at all in tubes 0, 1 and 2.

The experiment described above is a fairly simple one, requiring little sophisticated equipment. However, for the great majority of CCD work, which often involves several hundred transfers, the more elaborate automatic devices described earlier must be used. In essence, however, the principle and results are exactly the same.

Use of CCD machines

The use of a simple CCD machine to fractionate tRNA is shown in the next example, as described by Doctor, Apgar and Holley (1961), who used a 200-tube apparatus, each tube of which held 10ml. of each phase. Three basic experiments were carried out: (*a*) 200 transfer CCD of yeast tRNA at pH 8; (*b*) 200 transfer CCD of yeast tRNA at pH 6 and (*c*) 400 transfer CCD of yeast tRNA at pH 6.

The solvent system used in (*a*) consisted of phosphate buffer, approximately pH 8, prepared by dissolving 1090g. K_2HPO_4 and 421·7g. KH_2PO_4 in water to a final volume of 4l. To this was added 1240ml. formamide and 1170ml. isopropanol, and the whole mixture shaken. The solvent systems in (*b*) and (*c*) were very similar, in that phosphate buffer, pH 6, containing $MgCl_2$ was prepared by dissolving 555g. K_2HPO_4 and 870g. NaH_2PO_4 in 0·001 M $MgCl_2$ to give 5l. In (*c*) this was supplemented with 500ml. formamide and 2200ml. isopropanol and shaken (ratio of volumes, upper to lower phase, about 2 : 1). In (*b*) 40ml. of isopropanol were used per 100ml. of phosphate buffer. Yeast tRNA (34mg.) was used and the time of the experiment was about 26h. The fractions, after neutralization with HCl (2N) were determined for absorbance at 260mμ, and every fifth fraction between 40 and 170 analysed (after the dialysis treatment described above) for amino acid-incorporation activity. The results obtained (Doctor, Apgar and Holley, 1961) were then compared with those of a 200 transfer CCD of 100mg. yeast tRNA at pH 6 and are shown in Figures 4.4 and 4.5. The separation of threonine and tyrosine activities by the CCD process was shown to be greater at pH 6 than at pH 8: the separation at pH 6 occurs within fifty tubes, whereas at pH 8 only about thirty tubes are required. Also, the recovery of activity is greater at pH 6 (80 per cent) than at pH 8 (60 per cent). For this reason, the 400 transfer CCD of the same tRNA was carried out at pH 6. The solvent system in this experiment was shaken and maintained at 23°C for a few hours. The entire apparatus was then treated with 10ml. of the lower phase, and the first twenty-five tubes with upper phase. Yeast tRNA (600mg.) was dissolved in 50ml. phosphate buffer equilibrated to 23°C,

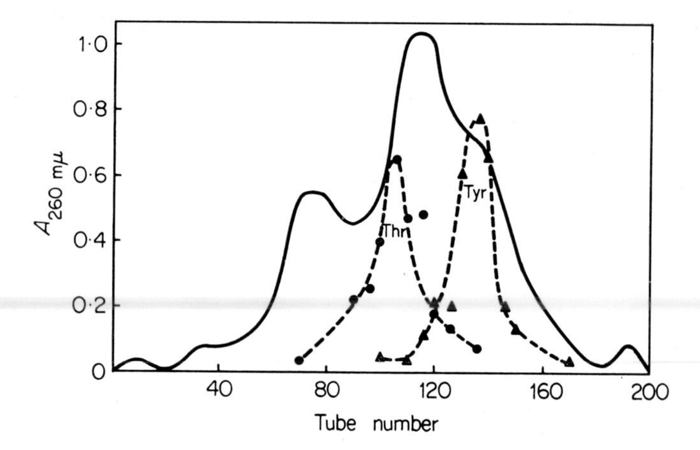

Figure 4.4. 200-Transfer countercurrent distribution of yeast amino acid-acceptor RNA's at pH 8. —, absorbancy (A) at 260 mμ; ●—●, threonine–acceptor activity; ▲—▲, tyrosine–acceptor activity. (After Doctor, Apgar and Holley (1961))

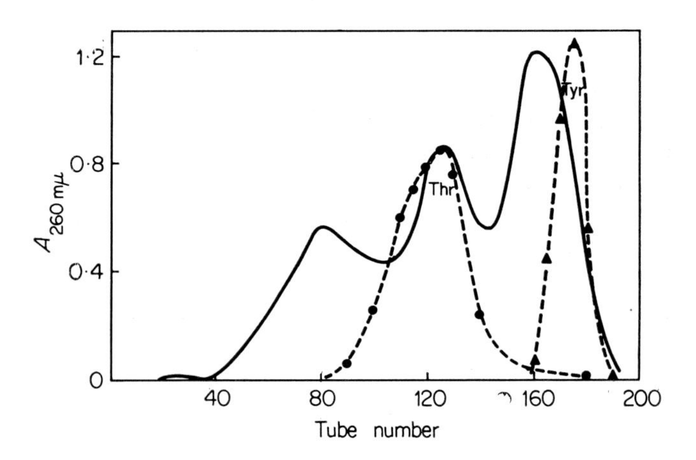

Figure 4.5. 200-Transfer countercurrent distribution of yeast amino acid-acceptor RNA's at pH 6. —, absorbancy (A) at 260 mμ; ●—●, threonine–acceptor activity; ▲—▲ tyrosine–acceptor activity. (After Doctor, Apgar and Holley (1961))

and appropriate amounts of formamide and isopropanol added. This was then made up to 50ml. of each phase by the addition of extra upper and lower preprepared phase. The resulting RNA solution was then used to replace the solvents in tubes 1 to 5 (i.e. about 100mg./tube loading concentration was used), and the apparatus set in motion for 200 transfers (24h.): After this time, tubes 101 to 200 were emptied and the tubes refilled with fresh solvents. The CCD process was then repeated for 200 more transfers by recycling the material in the apparatus (20h.). In order to obtain the pattern given by tubes 101 to 200, a second identical distribution was carried out to the 200 transfer stage, and tubes 1 to 100 emptied and replaced with fresh solvents. The recycling of the material was then carried on for 200 more transfers. The absorbancies of each fraction at 260mμ were measured from both sets of results and, after dialysis, each fraction was assayed for six amino acid-accepting activities as shown in Figure 4.6. Valine- and tyrosine-tRNA's were purified about ten-fold, whereas the threonine and leucine activity curves showed

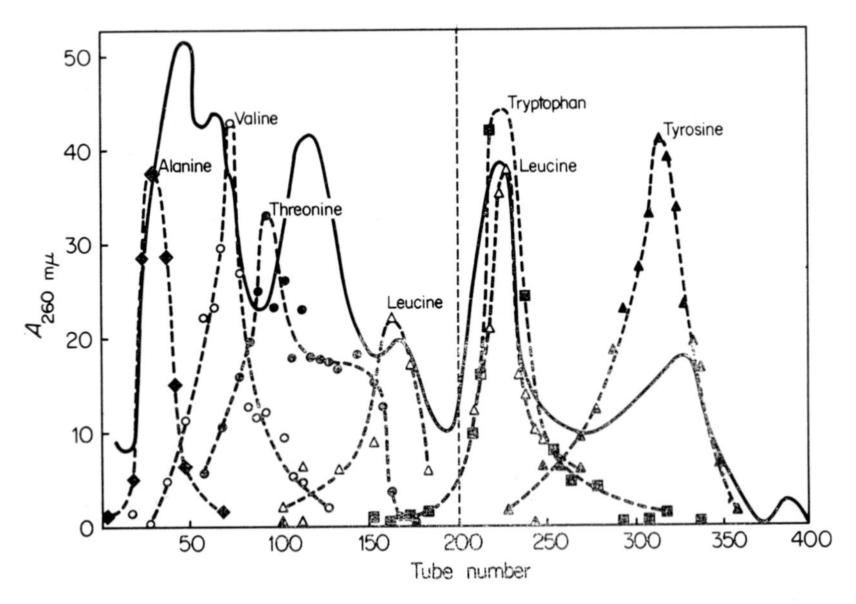

Figure 4.6. 400-Transfer countercurrent distribution of 600 mg. of yeast amino acid-acceptor RNA's at pH 6. —, absorbancy (A) at 260 mμ; ◆—◆, alanine–acceptor activity; ○—○, valine–acceptor activity; ●—●, threonine–acceptor activity; △—△, leucine–acceptor activity; ■—■, tryptophan–acceptor activity; ▲—▲, tyrosine–acceptor activity. (After Doctor, Apgar and Holley (1961))

definite heterogeneity, and leucine activity was resolved into two peaks. Evidence from redistribution of the material recovered from the two sides of the threonine peak indicated that this separation was real, and not an artifact arising from the composite 400 transfer distribution curve.

The heterogeneity of tRNA's specific for one amino acid has also been demonstrated by other workers using, in some cases, different systems. Goldstein, Bennett and Craig (1964) performed CCD of 970 transfers to separate sixteen tRNA's of *E. coli.* Using 1·7 M phosphate buffer pH6, formamide and isopropanol (10:0·75:3·5, v/v) they found twenty-nine specific tRNA's, six of which had one type of tRNA molecule per amino acid, eight had two types of tRNA molecule per amino acid, proline had three types of tRNA and leucine had four tRNA's.

Zachau and coworkers (1961) using their modified system in which the tRNA molecules were in the form of the tri-n-butylamine salt, have also investigated tRNA fractionation by CCD. In particular, they carried out a large scale isolation using CCD of serine-tRNA (31 μ-moles per mg. RNA,) which was contaminated with less than 1 per cent of other activities. A second CCD separated this serine-tRNA into two distinct forms (Karau and Zachau, 1964). In a subsequent investigation, Thiebe and Zachau (1965) tested sixteen amino acids for tRNA acceptability, and, using 200 transfers in their tri-n-butylamine system, obtained similar results to other workers in this field. Morever, the two phenylalanine- and three valine- tRNA's separated by the CCD were also tested on MAK columns (see Chapter 6) and found to be present in unfractionated tRNA samples, indicating that they were indeed separate native species, and not artifacts produced by the CCD. It was also found that the extent of stimulation of the phenylalanyl-tRNA by poly U in an *E. coli* ribosomal system was different for the two types of phenylalanyl-tRNA isolated.

The system devised by Kirby (1960) consisted of 2·5 M phosphate buffer, water and an organic solvent (1 : 1 : 0·8, v/v) in which the organic solvent was 2-ethoxyethanol, 2-butoxyethanol and *N,N*-dibutylaminoethanol (100 : 50 : 7·5) and was used to fractionate rat liver RNA according to base composition. It was found that (Kirby, 1962a) the fractionation depended on the greater solubility of adenine-rich regions in the aqueous phase. However, if RNA was prepared by the phenol/8-hydroxyquinoline method (Kirby, 1962b) no fractionation could be achieved in this system. This difficulty was overcome by Kirby and coworkers (1962) by altering the organic phase so that, as the partition of tripentylammonium acetate was varied by the presence of 0·033 M potassium citrate, the RNA could be transferred from the organic to the aqueous phase. This resulted in a fractionation pattern similar to the previous one, in that guanine-rich RNA remained, while adenine-rich RNA moved with the mobile phase.

This system was compared to the phosphate/formamide/isopropanol system for tRNA fractionating properties by Apgar, Holley and Merrill (1961), and shown to be essentially similar, while the modification of Doctor and Connelly (1961) only resulted in slightly better separation of valine- and threonine-tRNA's at the expense of the separation between alanine- and valine-tRNA's.

The sequence analysis of tRNA molecules required larger quantities of purified, specific tRNA molecules than were currently available at the time. For this reason, Holley and coworkers (1963) scaled up the CCD fractionation methods so that larger quantities of purified material could be isolated. The solvent system was altered slightly (2200g. K_2HPO_4, 3400g. NaH_2PO_4 in 16l. of distilled water, formamide 1200ml. and isopropanol, 5200ml.), and a larger apparatus capable of holding 40ml. per phase was used. Yeast tRNA (4g.) could then be fractionated in a very similar manner to that already described. Initially 200 transfers were carried out, and the various fractions assayed for alanine-, valine- and tyrosine-accepting activity. In this case, the dialysis and concentration techniques were not used, but aliquots of each tube were diluted with water and passed through Sephadex columns. This desalted the fractions and allowed the RNA to be assayed directly. The most active fractions for each amino acid were then recovered, dialysed and concentrated by extraction into, and from, methyl cellosolve. The whole procedure took one week. The partially purified fractions were than redistributed twice for 800 transfers at 24°C using the same solvent system. This was adequate for the alanine- and valine-tRNA's, but greater purification of the tyrosine-tRNA was obtained in the second distribution by slightly altering the conditions. The isopropanol in the system was increased (from 5200 to 6150ml.) and the fractionation carried out at a slightly higher temperature (26°C), which in effect lowers the partition coefficient. This procedure provided the starting materials for elucidation of the sequences of these three tRNA's. The nucleotide composition of the three tRNA's was determined by digesting the molecules with either alkali or pancreatic RNase (Holley and coworkers, 1963). These digests established that there were significant differences, and that small amounts of unusual bases were present in each. For example, valine-tRNA contains methylated adenylic acid but not methylated guanylic acid.

The nucleotide sequences of two serine-tRNA components has also been investigated (Rushizky and coworkers, 1965) but no major variations between the molecules could be seen. Thus the properties which cause variation in single amino acid-tRNA's cannot be attributed entirely to differences in nucleotide sequence. These types of studies, together with techniques in which isolated oligonucleotides were compared for overlapping sequences, finally resulted in the determination of the first complete nucleotide sequence of any nucleic acid molecule. This was alanine-tRNA isolated from yeast by Holley and coworkers (1965).

DNA fractionation

The fractionation of DNA by CCD has not been extensively investigated. Kidson and Kirby (1963) using a tri-n-amylammonium acetate and tri-lithium citrate system in 80 transfers, were able to show that heat-denatured *E. coli* DNA with a high GC content remains at the aqueous end, while AT-rich DNA travelled in the organic phase. These partitions were extremely sensitive to the lithium citrate concentration. The system was later extended to the fractionation of DNA from mammalian and bacterial sources. The degree of resolution depended on the amount of partial strand separation between the molecules, and evidence from T_m studies on the fractions, together with the reaction with formaldehyde, and the buoyant density, showed that base compositional differences were not involved, and that there were no S-value differences (Kidson and Kirby, 1964).

Partition column chromatography

The countercurrent distribution systems outlined above have proved very valuable in the investigation of tRNA. There are, however, practical difficulties involved in operating the equipment. The cost, by modern standards, is not excessively high but the average machine is large, and for the best results the temperature of operation must be kept constant. Cleaning and servicing 200 or more glass decantation units is tedious, and the volumes of buffers, etc., large. The technique has therefore not been very widespread in its application. The principles are, however, valuable, and it is the practical difficulties involved in achieving discrete 'plates', i.e. tubes, which complicate the method.

A continuous partition between two liquids would not have the advantages of discrete 'steps', and the calculation of theoretical curves would be more difficult, but it would be possible to evolve simpler equipment and handling techniques. Early attempts to do this simply involved the introduction of a less dense liquid at the bottom of a column of a more dense immiscible liquid, the former being allowed to rise through the latter as small droplets. After partition of a solute between the two liquids, the 'upper' layer was removed to a second column containing the 'lower' phase as shown in Figure 4.7. This was repeated for the required number of times but the method was not strictly continuous, as the 'stationary' phase was fragmented. In order for the system to be truly continuous, one of the liquid phases must be immobilized, in a similar way to the reacting groups on the ion-exchange resins, in a manner which allows a large surface area of reaction.

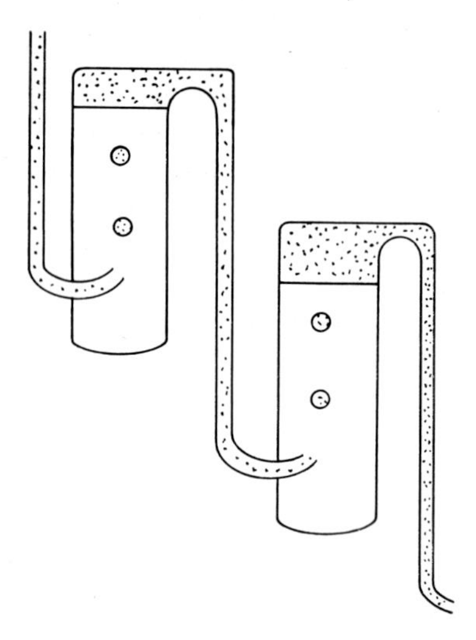

Figure 4.7. Partition of a molecule between two immiscible phases, one being the solvent containing the molecule which is introduced into the second more dense stationary phase in a column (see text)

tRNA fractionation

Tanaka, Richards and Cantoni (1962) replaced the 'upper' phase by a column of Sephadex G-25 which is a hydrophilic substance and, using the 'lower' phase developed by Zachau and coworkers (1961), converted tRNA's into the tributylamine salt and partitioned them through the column. Separation of several tRNA species was observed, and a second fractionation of the serine-tRNA region was able to resolve the tyrosine- and serine-tRNA's more fully.

Transfer RNA from *E. coli* has also been resolved by partition columns composed of Sephadex G-25 beads by Muench and Berg (1966a). They used the biphasic solvent mixture composed of potassium phosphate buffer (pH 6·80), ethoxyethanol, butoxyethanol, mercaptoethanol and triethylamine. Linear variation in the triethylamine concentration resulted in an exponential variation in the partition coefficient of tRNA in the mixture, thus allowing tRNA, dissolved in the aqueous phase immobilized on the Sephadex beads, to be extracted by the mobile phase containing a gradient of triethylamine. The authors claimed that a twenty-four-fold enrichment could be obtained during one passage over the column.

Alternatively, a hydrophobic stationary component can be used, such as diatomaceous earth (Chromosorb, W), and 4 per cent w/v dimethyl-dilauryl ammonium chloride in isoamyl acetate used as the mobile phase (Kelmers, Novelli and Stulberg, 1965). Sodium chloride gradient elution fractionated sixteen tRNA species, and multiple forms of leucine-, arginine- and serine-tRNA's were observed. The Mg^{2+} ion concentration was important for good results, and phenylalanine-acceptor tRNA could be concentrated twenty-nine-fold with 74 per cent purity. Kelmers (1966) extended this finding to the preparation of highly purified tRNA which would accept phenylalanine.

DNA fractionation

DNA has also been fractionated on partition columns. Kidson (1969) has developed a column in which the immobile phase is hydrophobic methylated Sephadex (Sephadex LH-20), and a salt gradient of low lithium ion concentration is used to elute the DNA. The solvent systems employed were adapted from the CCD system of Kidson and Kirby (1963), and consisted of *t*-amyl alcohol, 2-methoxyethanol, 2-butoxyethanol and tripentylamine in the following proportions:

(*a*) *t*-amyl alcohol–2-butoxyethanol–2-methoxyethanol (5 : 4 : 1, v/v, 28 volumes)

(*b*) tripentylamine–glacial acetic acid–solution (*a*) (6 : 1·08 : 100, v/v, 21 volumes)

(*c*) 0·033 M trilithium citrate in water (52 volumes)

The amounts of water and trilithium citrate were varied to give a range of trilithium citrate concentration of 2·5 mM to 0·7 M in the 52 volumes of the total volumes of upper and lower phases. The mixtures were shaken and allowed to separate in the dark.

Sephadex LH-20 was first equilibrated with the organic phase of the solvent mixture, and packed under atmospheric pressure into columns (20 × 1cm.). The organic phase in the external volume of the column was then replaced with the aqueous phase, all at the starting concentration of trilithium citrate. The DNA solution (5 to 100 μg.) was applied to the column in 1ml. of 2·5 mM trilithium citrate, and the column washed with 5ml. of the same buffer, and developed by a linear gradient from 2·5 mM to 10 mM trilithium citrate (total volume 130ml.). Usually about 70 1·9ml. fractions were collected for assay. The organic solvents were extracted with diethyl ether, and the DNA solutions dialysed against the required buffer. The fractionation of DNA extracted from *E. coli* (Kidson, 1969) is shown in Figure 4.8. DNA began to elute at about 4 mM trilithium citrate, and was completely eluted by 7–8 mM trilithium citrate. By altering the slope and extent of the ionic gradient, the elution

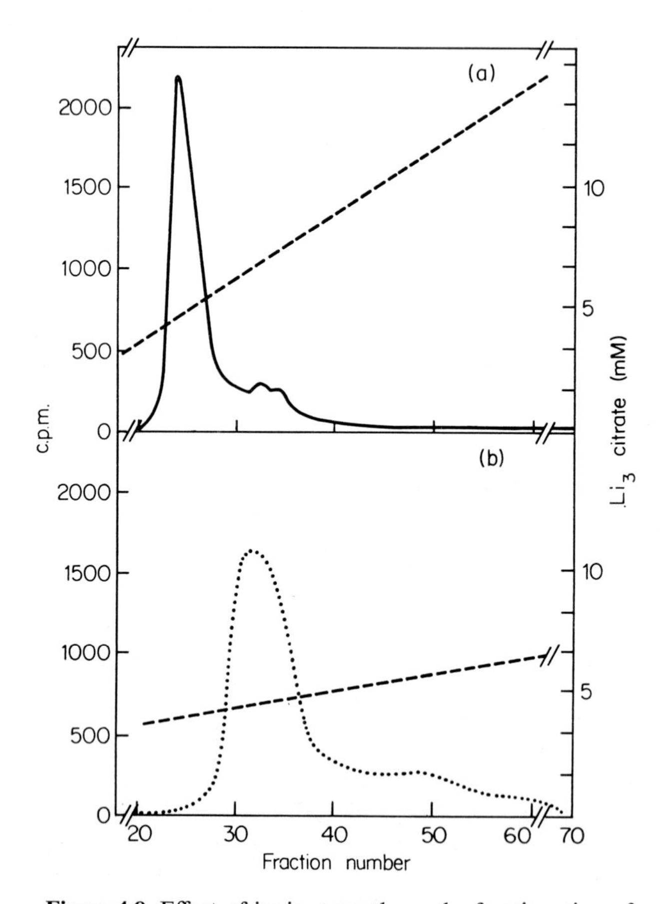

Figure 4.8. Effect of ionic strength on the fractionation of ^{3}H-labelled native *E. coli* B DNA. (a) Gradient from 2·5 to 17 mM trilithium citrate, total elution volume 130 ml. (b) Gradient from 2·5 to 8 mM trilithium citrate, total elution volume 130 ml. (After Kidson (1969) *Biochem.* **8,** 4376. Copyright 1969 by the American Chemical Society. Reprinted by permission of the copyright holder)

profile could be altered, and initiation of the gradient at higher ionic strengths (10 mM trilithium citrate) resulted in all the material eluting at the void volume of the column. The Li^+ concentration in the aqueous phase was the determining factor in the elution behaviour of double-stranded DNA at constant amine concentration. When the amine concentration in the organic phase was reduced, the DNA eluted at the void volume of the column, even at low ionic strengths. Most DNA preparations fractionated into a major peak and a series of minor peaks at higher ionic strengths. The reaction of the recovered fractions with ^{14}C-formaldehyde (Kidson, 1969) is shown in Figure 4.9 and indicates that there are regions

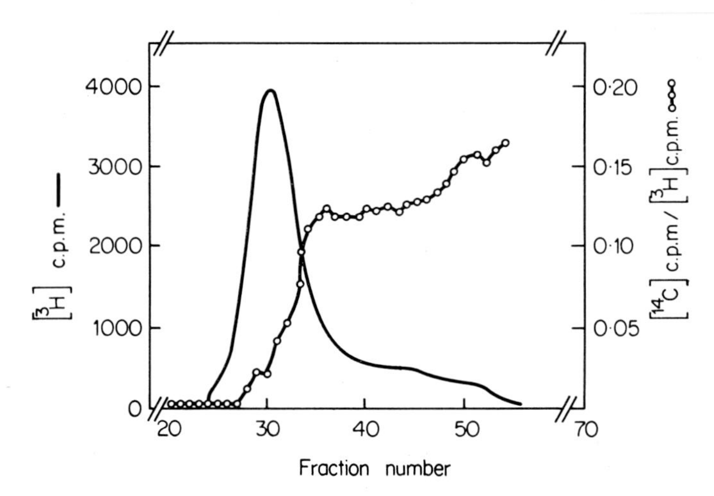

Figure 4.9. Reaction of [^{14}C] formaldehyde with recovered column fractions. Native ^3H-labelled XCl DNA was fractionated on a 20×1 cm. column. Individual tube fractions were recovered and treated with [^{14}C] formaldehyde. (After Kidson (1969) *Biochem.* **8,** 4376. Copyright 1969 by the American Chemical Society. Reprinted by permission of the copyright holder)

of single-strandedness in these minor peaks, although the degree of non-base paired regions cannot be very high since the hypochromicity values, derived from melting profiles, were not significantly affected. Completely heat-denatured DNA was retained on the partition columns, and could only be eluted at high trilithium citrate concentrations (10 mM to 0·5 M).

DNA molecules appear to be adsorbed on to the Sephadex LH-20 beads containing the organic solvent phase, by a mechanism which appears to depend on the interaction of amine with the phosphate groups of DNA in exchange for lithium ions (Kidson, 1969). This exchange is reversed at higher Li^+ concentrations, and the molecules consequently elute. Regions of unpaired bases react more fully with the organic phase in which they are soluble at low Li^+ concentration. Higher Li^+ concentrations are therefore required to alter the net partition coefficients in favour of the aqueous phase, and elute them from the column.

This ability of the partition columns to discriminate between molecules having slightly altered structure was utilized by Kidson (1968) during an investigation into the replication of *E. coli* DNA. Growing *E. coli* cells were 'pulsed' for very short periods of time with ^{3}H-TdR, and the resulting bacterial DNA fractionated on partition columns as previously described (Figure 4.10). The results obtained showed that, although some of the ^{3}H label was always associated with the bulk of the DNA, some label, at short pulses, was to be found in the regions of the profile in which a more 'open-structured' DNA had been demonstrated. The ^{3}H pulse-label could be chased with non-radioactive TdR until it coincided with the bulk DNA profile. This would indicate that newly synthesized DNA contains regions of open-strandedness or regions of complete single-strandedness near the replicating point.

Aqueous polymer two-phase systems

Theoretical considerations

The organic/aqueous systems discussed have been the most common in CCD work and in partition experiments generally. However, this is not the only type of immiscibility utilized in separation experiments. If a 0·72 per cent aqueous solution of methylcellulose is mixed with a 2·2 per cent aqueous solution of dextran, the mixture becomes cloudy, and if allowed to stand for a short time, separates into two layers both of liquid form. On analysis, the top layer consists of 98·96 per cent H_2O, 0·65 per cent methylcellulose and 0·39 per cent dextran, while the bottom layer contains 98·27 per cent H_2O, 0·15 per cent methylcellulose and 1·58 per cent dextran. In other words, the bottom layer contains most of the dextran whereas the top layer is richer in methylcellulose. Repeated shaking and standing of this liquid always results in the same two liquid layers, which can therefore be said to be immiscible and in equilibrium.

A detailed study of many such systems utilizing ionic and non-ionic polymers showed similar findings, in that one polymer collected in one phase, while the other polymer concentrated in the other phase. Liquid-phase systems may also be formed between a polymer and two low molecular weight compounds such as polyethylene glycol, potassium phosphate and water, in the same way that an aqueous solution of a polyelectrolyte and a salt form a liquid two-phase system. For example, a dextran sulphate solution can be made to separate into two phases by the addition KCl or CsCl.

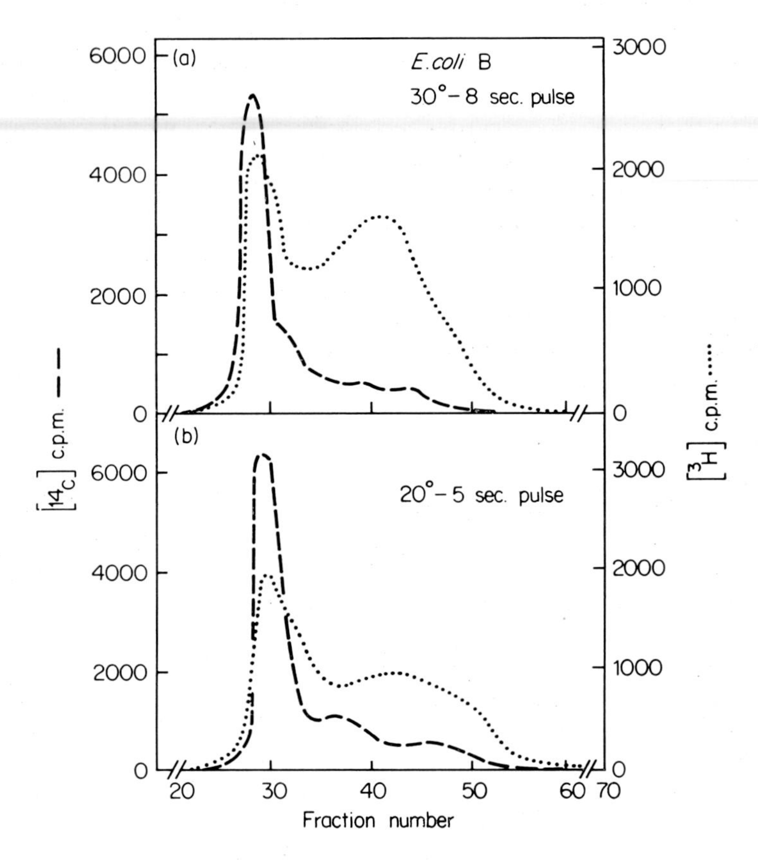

Figure 4.10. Partition column fractionation of DNA from *E. coli* B pulsed in logarithmic phase (a) for 8 sec. at 30°C and (b) for 5 sec. at 20°C, using ^{14}C-TdR for two generations as the general label and ^{3}H-TdR as the pulse-label. (After Kidson (1968))

Albertsson (1960) has classified the types of results obtained when mixtures of two different polymer solutions come together:

(*a*) 'Complete miscibility': a homogeneous solution is obtained (an event which does not often occur).

(*b*) 'Incompatibility': two phases form, each of which contains the major proportion of one of the polymers.

(*c*) 'Complex coacervation': two phases form but both polymers collect in only one of the phases.

Two factors determine which of the three possibilities above actually occurs when polymer solutions are mixed. One is the entropy gain which occurs when molecules are mixed, and the second is the interaction which takes place between the molecules. The entropy gain is related to the number of molecules taking part in the mixing process and therefore, on a molar basis, is approximately the same for large and small molecules. This is not true in the case of the interaction energy between molecules, as this increases with the size of the molecules, being related to the sum of all the interactions between each small segment of the molecules. In the case of very large molecules, therefore, the interaction energy/mole will tend to dominate the mixing entropy/mole, and this in turn will determine the result of mixing the two polymers.

Case (*a*) 'complete miscibility', would only occur when there was almost a complete absence of attractive or repulsive forces, which is seldom observed in practice. Case (*b*) 'incompatibility' occurs when the interacting forces are repulsive in character, so that the most energetically favoured state is when the two polymers are separated, and each is surrounded by its own type of molecule rather than being mixed. This is the most common result obtained from mixing two polymer solutions, and can occur between non-ionic polymers, polyelectrolytes or both. The final case (*c*), arises when there is a strong attractive interaction between the two polymers. This causes a tendency for the two to come together and collect in a common phase. The forces necessary to accomplish this must be great, however, but mixing is possible between oppositely charged polyelectrolytes for example.

In an aqueous mixture of two polymers, a phase separation will only take place if the constituents are present in a defined range of proportions. These compositions, at which phase separation occurs, can be represented graphically, in the form of phase diagrams. Figure 4.11 shows a theoretical phase diagram obtained by mixing polymers A and B in water. The concentrations of the two polymers are plotted on the abscissa and ordinate and are expressed as percentages. The curved line dividing the two areas is known as the binodial, and all the mixtures of polymers which have compositions shown as points above the line will separate into two phases, while those polymer mixtures which have compositions shown as points below the line will not. For example, the mixture represented by point X, will give a two-phase system, but that shown as point Y will not, and will remain as an homogeneous solution.

Consider a mixture of polymers represented by point X. Since it is above the binodial line, it will separate out into two phases, each phase of which will also have a definite composition. These compositions can then be represented by points on the binodial line as shown in Figure 4.12.

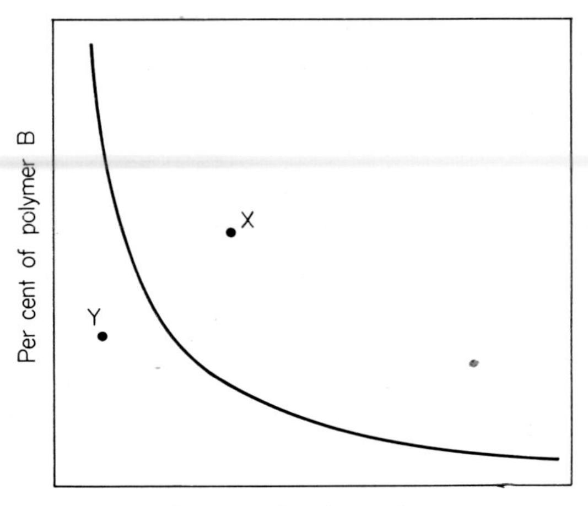

Figure 4.11. Phase diagram of a mixture of polymers A and B. The line separates the two areas at which phase separation takes place

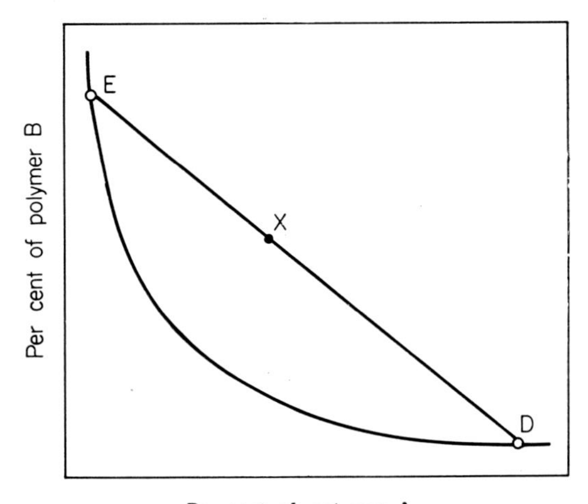

Figure 4.12. Phase diagram of mixtures of two polymers A and B showing a tie line representing the mixture at point X

Pairs of points (i.e. E and D) are called nodes, and the lines joining them, tie lines. Point X represents the total composition and lies on the tie line. Thus any of the total compositions on this tie line will also separate out into phases having the compositions of points E and D, but with different volumes of the two phases. If the compositions are expressed in per cent w/w, then the ratio of the weights bottom/top phase is equal to the ratio between the lines XE and XD. Thus the weight of polymer A in the top phase (m_t) plus the weight of polymer A in the bottom phase (m_b) should be equal to the total weight (m) of the polymer A, i.e.

$$m_t + m_b = m \tag{1}$$

If V_t = volume in top phase, d_t = density in the top phase and C_t, C_b and C = concentration of polymer A in the top phase, bottom phase and total concentration (w/w per cent), respectively, then

$$m_t = V_t \cdot \frac{C_t}{100} \cdot d_t$$

and similarly

$$m_b = V_b \cdot \frac{C_b}{100} \cdot d_b$$

also

$$m = C \times (V_t d_t + V_b d_b)$$

by substitution in equation (1)

$$V_t.C_t.d_t + V_b.C_b.d_b = (V_t d_t + V_b d_b)C$$

which can be expressed as

$$\frac{V_t.d_t}{V_b.d_b} \quad \frac{C_b - C}{C - C_t}$$

as we have already seen $C_b - C$ can be deduced from XD and $C - C_t$ from XE, therefore:

$$\frac{V_t.d_t}{V_b\, d_b} = \frac{\text{XD}}{\text{XE}} \text{ or } \frac{V_t}{V_b} = \frac{d_b}{d_t} \cdot \frac{\text{XD}}{\text{XE}}$$

The densities of polymer phases in water are not very different from water itself (1–1·1) so that the volume ratio may be obtained, approximately, from the distance XD and XE on the tie line as shown in Figure 4.13. As the total composition X approaches the binodial line, the difference between the two phases becomes smaller and at a certain point, termed the critical point K, the volumes and compositions become theoretically equal. Around this point, very small changes in composition can result in a change from one phase to two phases with very nearly equal volumes.

Any point on the binodial line has a 'critical' composition in that small compositional changes give rise to marked phase transitions, but only at the critical point are the volumes of the two phases produced equal. Phase systems about the critical point require very careful experimental control and should whenever possible be avoided.

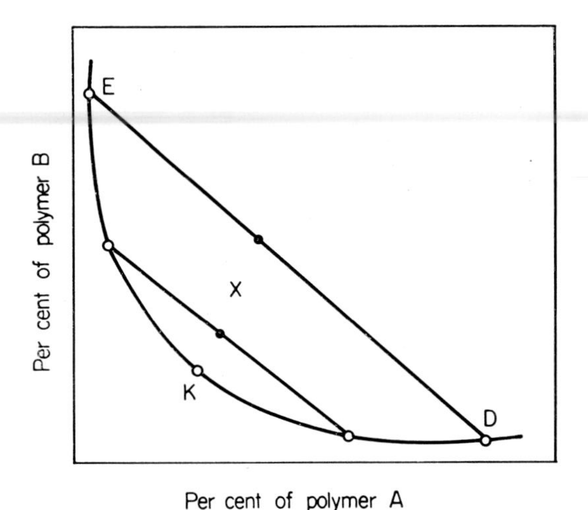

Figure 4.13. Phase diagram of polymers A and B showing varying composition approaching the critical point

Experimental construction of a phase diagram

The experimental determination of the binodial can be accomplished as follows (Albertsson, 1960). A known quantity of one polymer (A) is made into a concentrated solution in a test tube, and a second concentrated solution of the other polymer (B), of known concentration, is added in a slow dropwise fashion in known amounts. After the addition of a certain amount, what was previously an homogeneous mixture becomes turbid, and a two-phase system results. The composition of the mixture at this moment is noted. Water (1g.) is then added to the mixture, and the solution should again become clear. More of the solution of polymer B is again added in a dropwise fashion, until the turbidity and phase separation is noted once more. The composition of the mixture at this point is recorded, and the whole procedure carried out several more times. This method will give a series of compositions close to the binodial, which can be plotted into a graphical form by plotting the concentration of polymer A against that of B.

Unfortunately certain practical considerations may make the change from a clear to a turbid solution difficult to detect. When the polymers are polydisperse, on adding the second polymer, a very weak turbidity appears first, probably because only the larger molecules of the first polymer separate out into two phases, then, as more of the second polymer is added, more of the first polymer separates out, and so on. This makes this type of determination rather tedious for very polydisperse polymers.

Alternatively, the compositions of the phases of a number of different total compositions can be analysed. This gives two points and the central point on a tie line, which, if the end points are joined for several such systems, gives the binodial line once more. If the total composition is known, then only one of the phases need be analysed, as the ratio of the weights in the two phases can then be calculated, and the composition of the other phase obtained.

The critical point can be determined by trial and error, such that a mixture is made which can be converted from a one-phase to a two-phase system with equal volumes of the phases, by the addition of a single drop of one of the polymer solutions. Alternatively, if the method of tie lines has been used, and some of them approach the critical point, a line can be drawn through the mid points of these, and extrapolated to the binodial line. The point of intersection corresponds, approximately, to the critical point.

Compounds used in phase systems

The polymers used must be relatively inert, well defined and readily available. Also, from a practical point of view, the phases produced must not have inconveniently long separating times.

Dextran is one of the most common components used in these systems. Dextran fractions are characterized by their limiting viscosity number $[\eta]$, number average molecular weight $\overline{M}n$ and the weight average molecular weight $\overline{M}w$. From these the polydispersity of the fraction can be obtained from the quotient $\overline{M}w/\overline{M}n$. In experiments where a low u.v. absorbance is necessary, the dextran can be reduced with $NaBH_4$ (pH changes neutralized with HCl), and then precipitated out of solution using ethanol. Dextran treated in this way can be dissolved directly in cold water. Dextran sulphate, prepared as the sodium salt (NaDS) from dextran fractions with limiting viscosity numbers of 70ml./g. and 68ml./g., can be used without further purification.

Methylcellulose is supplied commercially as 'Methocel', and information regarding its data can be obtained from the *Methocel Handbook* published by Dow Chemical Company. The fractions MC4000, MC400, MC10 were the ones used by Albertsson (1960) who found that they were rather polydisperse. Solutions of methylcellulose were prepared by adding hot water (80 to 90°C) to the dry powder, and shaking vigorously for a few minutes to cause wetting. An equal volume of cold water was added, and the flask again shaken before being allowed to stand, with occasional stirring, until it reached room temperature. The powder swells and slowly dissolves, but should not be allowed to sediment to the bottom of the flask. Concentrations can be checked by dry weight.

Polyethylene glycol (PEG) is usually obtained in the form of 'Carbowax', and can be dissolved directly in water. Dry weight determinations for concentration are not very satisfactory for this compound, and freeze drying should be used. Any u.v. absorbing contaminants can be removed by precipitating the PEG from an acetone solution with ether.

Biological applications

The following are a few examples of how aqueous polymer systems may be used to isolate and/or concentrate biological particles and molecules (Albertsson, 1960). The standard preparations of microsomes obtained by various differential centrifugation techniques and similar methods of preparing ribonucleoprotein particles, are always contaminated with comparatively large vesicles, possibly arising from the endoplasmic reticulum. These vesicles, in the case of rat brain, are very variable in size, but generally have a lower density than the ribonucleoprotein particles. Thus inevitably, vesicles larger than the ribonucleoprotein particles will sediment in a very similar manner, and hence be difficult to separate by differential centrifugation. A polymer phase system of dextran and methylcellulose was therefore chosen (Albertsson, 1960; Albertsson, Hanzon and Toschi, 1959) as the basis for separation of ribonucleoprotein particles from rat brain microsomes. The microsomes are first distributed in a dextran and methylcellulose system in which the upper phase is 58ml., and the lower phase is 22ml. (total 80ml.), derived from 0·68 per cent (w/w) dextran, 0·36 per cent (w/w) methylcellulose and 98·96 per cent (w/w) water total composition. The partition coefficient of ribonucleoprotein (RC-) particles is such that the top phase will contain mostly RC-particles and no vesicles. In addition, because of their larger volume, a considerable amount of the ribonucleoprotein particles will be present in this phase. However, soluble substances, such as proteins, will also be present in large concentration in the top phase. The top phase is removed, and fresh top phase added to the old bottom phase, and more RC-particles extracted. The top phases are combined, and suitable amounts of dextran and methylcellulose added so that a new system is set up in which the total composition is 0·50 per cent (w/w) dextran, 1·20 per cent (w/w) methylcellulose and 98·30 per cent water. This is much further removed from the critical point of this system, and separates into two layers, the upper having a volume of 126ml. and the lower 14ml. As the RC- particles have a much lower K value in this new system, they concentrate into the bottom phase which, because of its smaller volume, has a concentrating effect. Soluble substances do not concentrate to such an extent. The RC- particles can then easily be removed by high-speed centrifugation.

The penultimate step in the above procedure resulted in a concentration of the particles into a relatively small volume of one phase. This effect can be very useful in isolation procedures which give very dilute solutions or suspensions at some point during the procedure. Consider the situation in which V_o is the volume of the original particle suspension and C_o is its concentration in particles per millilitre. A solution of the two polymers with volume V is now added, to give the required phase system, and after mixing and separating, the volume in the lower phase is V_b. Thus the volume in the top phase, V_t, is given by

$$V_t = V_o + V - V_b$$

If the partition coefficient of the particles is K ($K = C_t/C_b$), then the total number of particles in the top phase plus bottom phase should equal the total number of particles

$$V_t.C_t + V_b.C_b = V_o.C_o$$

The concentrating factor α is defined as

$\alpha = C_b/C_o$ and can be calculated from the previous two equations

$\alpha = V_o/V_b(1 + V_t \times K/V_b)$

From this equation it can be seen that for large values of α, K must be small and V should be as small as possible, so that V_t should not be too large compared with V_o.

The partition of nucleic acids in aqueous polymer two-phase systems has been investigated by Albertsson (1965a). Several factors were examined to see in what way they influenced the partition coefficient. These were variation of electectrolytes, polymer concentration and nucleic acid conformation. Using a buffer consisting of 0·005 M NaH_2PO_4 and 0·005 M Na_2HPO_4 and dextran/PEG, four systems were examined, and several different types of electrolyte and various concentrations studied. It was found that the partition coefficient was markedly dependent on the type and concentration of the electrolyte, and that often a small shift could move the nucleic acids completely from one phase into the other. Also, the transition from single-strandedness to double-strandedness caused a 100- to 1000-fold increase in the partition coefficient. This was obviously a useful property which could be applied further as a basis of separation.

Albertsson (1965b) developed a system in which phase separation occurred in 1 to 2 min. The apparatus consists basically of a partition cell block which is made up of two cylindrical plates, the stator and rotor. The stator upper surface has a shallow annular groove, and in this groove a number of shallow cavities which form the lower parts of the partition cells. The depth of these cavities is 2mm. The rotor rests in the groove of the stator, and can be rotated about its axis, being held in place by the inner and outer edges of the groove. The lower surface of the rotor also has cavities which exactly match the cavities in the stator. Thus, when these two sets of cavities are aligned, they form the upper and lower parts of the partition cells. The cells are filled with lower phase to the level of the stator surface, and then topped up with upper phase. The whole head is shaken to mix the phases together, and then allowed to stand so that the phases can separate. Because of the very thin layers the separation time, depending on the type of system used, is short and, when complete, the upper phase is moved on to the next lower phase by moving the rotor one step. This moves all the upper phases along in successive stages.

A circular sequence of cells in this way allows recycling procedures, so that the total number of transfers may be increased above the 60 or 120 cells per head, and hence allow an increased maximum separation. Each cell is provided with a hole in the rotor part so that the cells may be easily filled and emptied. These in turn can be brought in contact with fraction collector tubes and, after a run, the fractions can be easily transferred to the tubes by inverting the whole head.

This technique has been applied to the CCD of virus particles and viral RNA (Oberg, Albertsson and Philipson, 1965). The system used consisted of a lithium phosphate buffer (0·05 M pH 6·0) containing 2 mM $MgCl_2$ and 0·5 mM NaCl, together with 5 per cent dextran 500 and 4 per cent polyethylene glycol 6000. Sixty transfers in this system, where the top phase was 0·7ml., the bottom phase was 0·5ml., shaking time 15 sec. and separation time was 3 min., showed that the intact poliovirus only travelled a short distance from the loading tube 0, whereas the RNA moved about three times this distance (Figure 4.14). The poliovirus could be recovered with 10 to 60 per cent infectivity.

The large variation in partition coefficients between native and denatured DNA, discovered by Albertsson (1965b) for the dextran/PEG system was further investigated by Pettijohn (1967) who examined the partition coefficients of linear and twisted circular double-stranded DNA of Polyoma virus. The partition coefficient of twisted circular DNA was ten times less than linear DNA. Partial denaturation introduced regions of single-strandedness into the linear molecule, while leaving the twisted molecule unaffected. This allows the partially denatured linear molecule to be transferred from the upper PEG phase to the lower dextran phase. Alberts (1967) carried out a series of experiments, in which he was able to show a quantitative separation of single-stranded (and partially single-stranded) DNA from double-stranded DNA, in a single extraction.

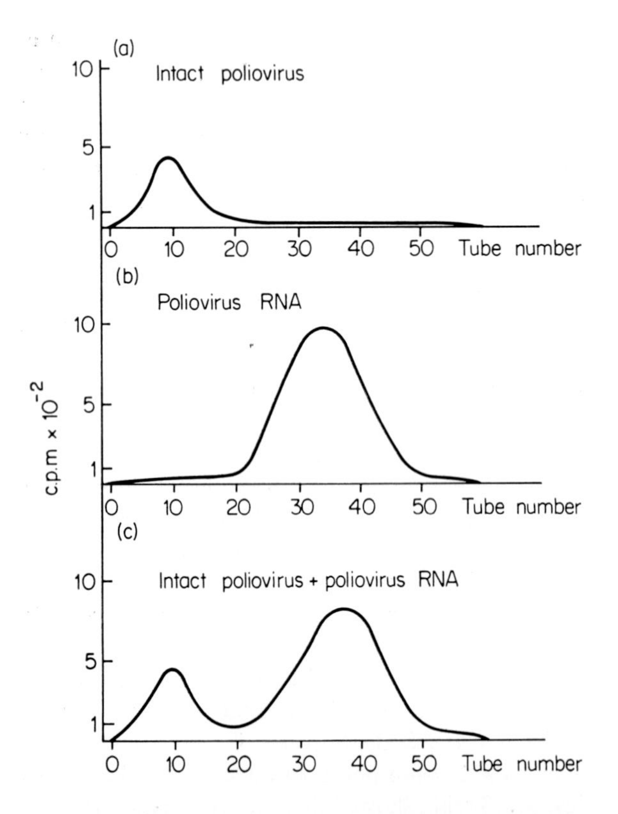

Figure 4.14. Countercurrent distribution in D_5^{500}-PEG_4^{6000}, 0·050 M lithium phosphate (pH 6·0), 2 mM $MgCl_2$ and 0·5 mM NaCl. The number of transfers was 60, shaking time 15 sec., separation time 3 min., top-phase volume 0·7 ml., bottom-phase volume 0·5 ml. and temperature 23°. (a) Intact poliovirus added to Chamber 0; (b) poliovirus RNA added to Chamber 0; (c) a mixture of poliovirus and poliovirus RNA added to Chambers 0 and I. (After Oberg, Albertsson and Philipson (1965))

Concentrated stocks of the phase system were kept frozen until required, and consisted of 16·8 per cent w/w Dextran 500 and 9·2 per cent PEG 6000 in distilled water. The stock was completely emulsified prior to use, and an aliquot of predetermined weight removed, and pipetted into a tared test tube. A known volume of a DNA solution (in 0·01 M sodium phosphate buffer, pH 6·8) was added, the mixture emulsified and the phases separated by a 5 min. low-speed centrifugation. The phases were withdrawn, diluted ten-fold and assayed for DNA. Figure 4.15 shows the effect of total polymer concentration on the partition coefficients of denatured and native DNA. With high concentrations of phase system, both native and denatured DNA favour the dextran-rich (lower) layer, and only at lower concentrations do the DNA's exhibit phase selectivity. Figure 4.16 shows the phases can be analysed for the type of DNA present by adopting the following technique. Solid CsCl can be added directly to the top (PEG-rich) phase containing native DNA and then centrifuged in the analytical ultracentrifuge. This procedure not only allows the direct CsCl gradient analysis of the DNA, but also removes the PEG, which is salted out of solution, and floats to the top of the liquid as a thin immiscible layer. It is not possible to examine the DNA in the bottom phase by the same direct addition of CsCl, as the dextran does not salt out, and seriously interferes with the gradient. If the pH of this phase is raised to 10 by the addition of NaOH, any DNA present can now be extracted into the new top phase (PEG-rich) and treated as before. If it is required to remove the PEG without first adding CsCl, a similar salting-out effect can be produced by adding solid potassium phosphate to a concentration of 23 per cent. Once the PEG layer has been removed, the DNA solution can be dialysed free of salt.

When the complementary strands of DNA are covalently linked together in some way, then the DNA acquires resistance to irreversible denaturation. This is because, since both strands can never be fully separated, their close proximity means that when the denaturing conditions are removed, the two strands can readily 'zip' back together again. Summers and Szybalski (1967) have utilized these crosslinked DNA molecules as a substrate for assaying radiation effects. If a single strand break is introduced into one of the chains of such a crosslinked molecule, then the 'zipping' mechanism will no longer function perfectly, as shown in Figure 4.17. This will result in two classes of new molecules after denaturation: a completely single-stranded fragment, and a bihelical segment (with the crosslink) having a single-stranded 'tail'. The Dextran/PEG system can distinguish between these two molecules and completely native molecules as shown previously. This transfer of DNA from the upper to lower phase, following denaturation, can be considered as a

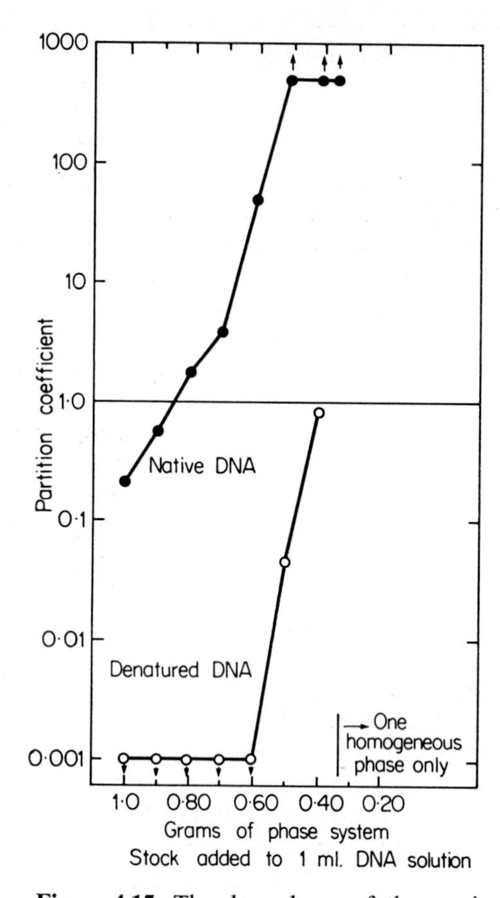

Figure 4.15. The dependence of the partition coefficient for native and denatured DNA on the proportion of phase-system stock added. Partition coefficients are expressed as top-phase absorbance divided by bottom-phase absorbance. Dialysed calf thymus DNA at 300 μg./ml. was either partitioned directly or denatured for 5 min. at 100° in the 0·01 M sodium phosphate buffer (pH 6.8) used for phase extraction. The concentrated polymer stock here consisted of 16·8 per cent dextran and 9·2 per cent polyethylene glycol. (After Alberts (1967) *Biochem.* **6,** 2527. Copyright 1967 by the American Chemical Society. Reprinted by permission of the copyright holder

measure of single-strand breakage in crosslinked molecules, and hence radiation effects can be assessed.

Summary

The techniques described in this chapter all depended on the partition of molecules between two immiscible phases. The CCD method using two liquid phases (an aqueous and an organic) has been successfully applied to fractionate tRNA into the various amino acid specific tRNA's as well as the multiple tRNA species of a single amino acid. There is a

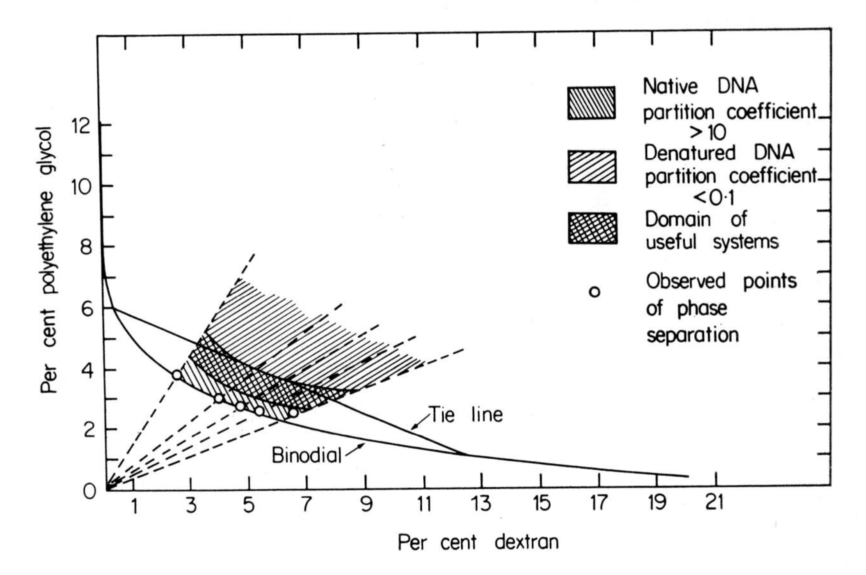

Figure 4.16. The phase diagram of a dextran–polyethylene glycol two-phase system and its relation to the separation of denatured from native DNA. Data such as that presented in Figure 4.21 have been used to determine the polymer concentrations in the phase diagram yielding partition coefficients greater than 10 for native DNA and less than 0·1 for denatured DNA. The region where both types of behaviour are observed has been arbitrarily defined as the domain of useful systems (cross-hatching). The curved 'binodial' indicates the minimum concentrations of dextran and polyethylene glycol required for phase separation, with all mixtures below this line consisting of one homogeneous phase. The 'tie line' shown is one of a system of almost parallel straight lines, each of which connects mixtures for which the individual compositions of top and bottom phases remain constant. (Along a given tie line, only the relative volumes of the two phases change and the composition of each phase is indicated by the two points of intersection of the tie line with the binodial). (After Alberts (1967) *Biochem.* **6,** 2527. Copyright 1967 by the American Chemical Society. Reprinted by permission of the copyright holder)

high percentage of recovery, and the method is suitable for large-scale isolation of tRNA for sequence analyses. The method is therefore comparable to ion-exchange chromatography in efficiency. However, the length of time required to achieve a good separation using CCD is greater than that for ion-exchange chromatography, due to the large number of transfers involved. The method is not applicable to DNA fractionation since it remains largely in the aqueous phase.

Partition chromatography using columns containing a stationary phase of either Sephadex G-25 or a Chromosorb W, provides an equally effective method for fractionating tRNA, and the resolution time is less than in the case of CCD.

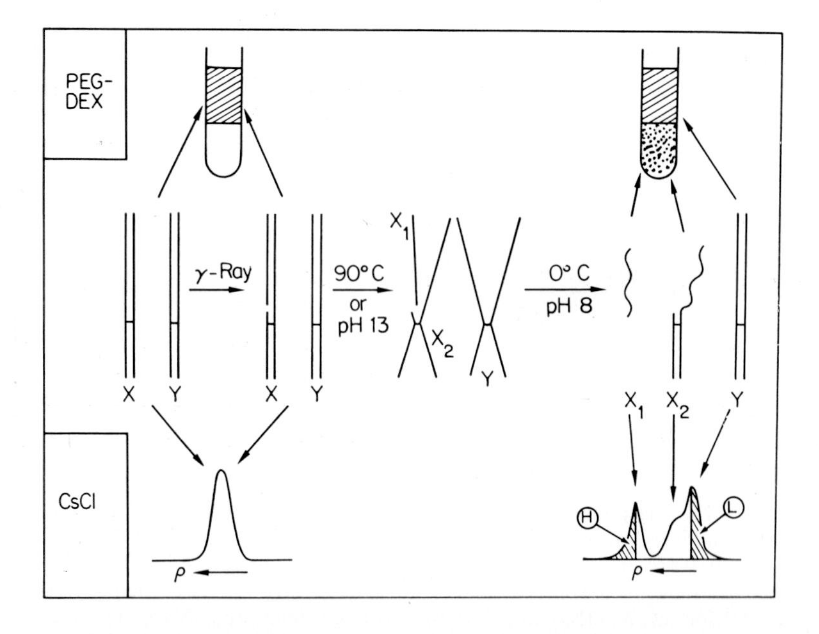

Figure 4.17. Schematic diagram illustrating the principle of the method for detection and quantitation of single-strand breaks in DNA.
Two crosslinked DNA molecules are represented in the centre row. The molecule labelled X sustains a single-strand break which results in the production of two fragments (X_1 and X_2) upon denaturation, while the unbroken molecule labelled Y perfectly zippers-up in its entirety. Perfectly bihelical (zippered-up) DNA can be separated from denatured and partially bihelical DNA by fractionation in a polyethylene glycol–dextran (PEG–DEX) two-phase system (upper row of the figure) or in a CsCl density-gradient (lower row of the figure). (After Summers and Szybalski (1967))

The careful choice of aqueous/organic phases and the adsorption of a methylated hydrophobic Sephadex LH-20 matrix allows the separation of native and denatured DNA. This is due to the fact that the interaction between the phosphate groups of DNA and the amino of the organic phase within the Sephadex beads, is stronger in the case of the denatured single-stranded DNA chains.

The use of aqueous/polymer two-phase systems such as dextran/PEG/ H_2O greatly increased the possibility of fractionating nucleic acids according to partition theory. These systems also provided an excellent method of isolating and concentrating a particular nucleic acid species, for, unlike CCD between two liquid phases, aqueous polymer two-phase systems have been more successfuly applied to DNA fractionation. Denatured DNA can be separated from native DNA in a single-stage procedure and in a very short time (5 min.). Moreover, the native DNA present in the PEG (upper) phase can be analysed immediately by CsCl density-gradient centrifugation without the time-consuming procedures of isolation and concentration. The denatured DNA in the dextran (lower) phase can also be transferred to a PEG phase by an appropriate pH change, and also be analysed by CsCl density-gradient centrifugation.

Aqueous/polymer two-phase systems have obvious potentialities with regard to separating circular DNA from linear DNA after alkali denaturation and neutralization. The circular DNA will renature and be extracted in the PEG phase, whereas the linear DNA will be irreversibly denatured, and remain in the dextran phase. Mitochondrial DNA could be separated from cytoplasmic DNA, R-factor from host chromosomal DNA and bacteriophage DNA from host bacterial DNA. The one-step procedure could supersede hydroxyapatite column chromatography (see Chapter 5) and the nitrocellulose technique (see Chapter 2).

CHAPTER 5

Molecular sieving, acrylamide gel electrophoresis and hydroxyapatite

Introduction

The methods described in this chapter have been used to fractionate nucleic acids in a way that depends on the chemical and physical properties of the molecules, the most important of which are structure and electrical charge. The obvious differences in physical structure between, for example, tRNA and 25S native DNA, allow their separation by a method which depends on molecular size. However, differences in secondary structure also exist, which can be utilized to provide a more effective separation.

The use of the inert dextrans in ion-exchange chromatography has already been described (Chapter 3), but the earliest use of dextrans *per se* was by Porath and Flodin (1959) who demonstrated their action as a kind of molecular sieve, the principle being that smaller molecules will be retarded because they have the chance of penetrating the gel particles, whereas the larger molecules will be excluded from the gel, and elute more quickly.

The negatively charged phosphate groups which form the backbone of the polynucleotide chain allow electrophoretic methods to be used for the separation of nucleic acids. Polyacrylamide gel electrophoresis provides a useful, if somewhat analytical, tool for RNA separations. This system 'drags' polynucleotides through a type of filter by using an electrical field as a force. The electrical charge on the molecules is important, but the structure and size of the molecules also contribute to their fractionation.

One method which fractionates according to secondary structure employs a column of calcium phosphate, which is both the matrix and the charge-carrying exchanger, to 'trap' nucleic acids out of solution in a positive fashion. The resulting complexes between the matrix and the nucleic acids are broken by increasing the phosphate concentration of the eluting buffer and the molecules selectively released. This method, although limited to a certain extent, has been used effectively for the separation of native double- and single-stranded DNA, which elute at differing phosphate molarities. Unfortunately the artifacts occasionally produced by certain types of fractionation, and the observed breakdown of high molecular weight RNA species, have prevented a wider use of the method, but its value in a more restricted way should not be overlooked.

Gel filtration or molecular sieving

Introduction

This technique fractionates chiefly according to differences in the molecular size (which in turn often depends on differences in molecular weight) but conformation is also important. The mechanism of gel filtration appears to depend largely upon the physical barriers which prevent the completely random distribution of certain solute molecules, the selectivity being a function of the molecular dimensions of the solutes, and the probability of their being able to pass through a hole or 'pore' of a given size. The classical example of this principle is the semipermeable membrane which is usually composed of cellulose sheets, and used for dialysis. These membranes only allow the passage of small molecules, and are generally used to desalt aqueous solutions of macromolecules. A sieving effect occurs, and the molecules are free to migrate at random. Initially the concentration of all the solute components is highest on one side of the membrane, so that the movement of those molecules that are diffusible is unidirectional. Eventually an equilibrium is reached, in which the concentration of the diffusible molecules is the same on both sides of the membrane and net movement ceases. It can be seen that the larger the 'external' volumes (i.e. outside the dialysis bag), the lower the concentration of the smaller molecules will be in the 'internal' volume (i.e. inside the dialysis bag). For this reason, repeated dialysis is necessary in order to ensure the complete removal of the small molecules.

The application of this phenomenon as a column separation method required the development of a substance which could act as if it were composed of hundreds of tiny dialysis bags, and be packed into a column. A gel-like substance was required which had a reticulate structure of definite pore size and 'sieved' the molecules as they moved unidirectionally. This has led to the development of gels of crosslinked polymers, in which the degree of crosslinking can be varied and carefully controlled to allow a defined pore size.

By definition, gels consist of two components, the dispersed substance and the dispersing agent, i.e. the gel-forming material and the solvent, respectively. Each component penetrates and stabilizes the other so that, within the system, a particle may move to any location without ever leaving the system. This can be shown in the case of true gels by the fact that small molecules diffuse with almost the same velocity as that which would obtain in free solution. In practice, gels are formed by long-chain macromolecules which are held together in the gel structure by junctions, or regions of junction points, the forces of which can be ionic, covalent, hydrogen-bonding or result from dipole–dipole interactions. Unlike dialysis, the separation of components achieved by unidirectional movement through homogeneous gels cannot be completely attributed to diffusion.

Dextran gels

Gel filtration chromatography of nucleic acids requires gel granules capable of discriminating between very high molecular weight substances, and hence requires pore sizes larger than those used for the separation of small proteins or polypeptide chains. In recent years, two types of granular gels have been developed which are now readily available commercially. The commercial preparations are excellent in quality, and have a very wide range of pore and granule size. For this reason it is not usually necessary for the research worker to prepare his own gels. However, it is interesting to at least know the principles behind their preparation.

It is considered important that the matrix of the gel be inert so that no undesirable secondary forces will disturb the principles outlined. For this reason, the forces holding the junctions together should not be ionic, and the polymer should not in turn contain any ionic groups. These requirements are met by a high polymer carbohydrate produced by *Leuconostoc mesenteriodes* when grown on sucrose, and called dextran. This substance consists entirely of glucose residues joined together by 90 per cent α-1-6-glycosidic linkages and, because of the three free hydroxyl groups per residue, is water soluble. The crosslinks between chains are introduced by means of epichlorohydrin which reacts with two hydroxyl groups on two different chains. The dextran solution is kept alkaline, and the reaction takes place exothermally as shown in formulae **1**

$$(a)\ \text{sugar—OH} + \overset{\overset{\text{O}}{\diagup\ \diagdown}}{\text{CH}_2\text{—CH}}\text{—CH}_2\text{Cl} \rightarrow \text{sugar—O—CH}_2\text{—CH(OH)—CH}_2\text{Cl}$$

$$(b)\ \text{sugar—O—CH}_2\text{—CH(OH)—CH}_2\text{Cl} + \text{NaOH} \rightarrow \text{sugar—O—CH}_2\text{—}\overset{\overset{\text{O}}{\diagup\ \diagdown}}{\text{CH—CH}_2} + \text{NaCl} + \text{H}_2\text{O}$$

$$(c)\ \text{sugar—O—CH}_2\text{—}\overset{\overset{\text{O}}{\diagup\ \diagdown}}{\text{CH—CH}_2} + \text{HO—sugar} \rightarrow \text{sugar—O—CH}_2\text{—CH(OH)—CH}_2\text{—O—sugar}$$

(1)

A number of side reactions can also take place, for instance the epichlorohydrin may be hydrolysed by the aqueous alkaline solution, and the chlorine released, or hydrolysis may occur after the epoxide has reacted with one chain, and hence prevent the formation of a crosslink.

Sephadex gels take up water and swell, a property that is dependent on the dextran concentration, its molecular weight and the ratio of dextran to epichlorohydrin. In general, the swelling properties of the gel increase with decreasing concentrations of the dextrans, lower dextran molecular weight and decreasing quantities of epichlorohydrin. They are classified according to their swelling properties, and the range of fractionation is determined by the corresponding porosity: for example, 'Sephadex G-200' regains 20·0 ($\pm$ 2·0) ml. of water per gram of dry powder, and gives a gel bed volume of 30 to 40ml./g. This has an approximate separation range (estimated by using peptides and globular proteins of known molecular weight) of 5,000 to 800,000 and is therefore capable of retarding tRNA molecules. It is important that the gel be completely swollen before use, and should be allowed to swell in excess solvent for the recommended periods of time for each type of gel, before packing into columns. 'Sephadex G-200' requires three days at room temperature or 5h. in a water bath at 100°C. The gels are stable to alkaline solutions and weak acids, but strong mineral acids will cause some hydrolysis of the glycosidic links, and strong oxidizing agents should not be used because they will affect the dextran. Swollen Sephadex gels may be dried after use for subsequent storage, by washing with water to remove salts, and then treated with 50 per cent ethanol. The gel shrinks to half its original volume and the remaining water may be removed by standing for half an hour in 99 per cent ethanol with regular shaking. Repetition of this process is often necessary. A free-flowing powder can then be obtained by drying at 60 to 80°C. Sephadex may be sterilized in the swollen state by autoclaving at 110°C for 40 min. without damage to its fractionating properties.

Alternative solvents to water may be used for all the G-types of Sephadex, and the beads are capable of swelling in dimethyl sulphoxide, formamide and ethylene glycol. The properties are then slightly different with respect to the degree of swelling.

Sephadex gel matrices substituted with ionic groups, for example DEAE-Sephadex, have been used successfully to fractionate nucleic acids (in particular tRNA) and were described in Chapter 3.

Agar and agarose gels

Agar can be extracted from red seaweed in a partially purified form and is a well-known material to bacteriologists. It has been shown by Araki (1956) to consist of a neutral component, agarose, which makes up the bulk of the material, and agaropectin which it is claimed contains all the ionic groups, e.g. carboxyl and sulphate. Agarose was described as a linear polysaccharide of D-galactose and 3-5-anhydro-L-galactose.

Agar can be used directly as a gel by dissolving in hot water and allowing the resulting solution to cool. This results in a very rigid gel, the degree of rigidity depending on the agar concentration. The acidic groups on the agaropectin cause pronounced electroendosmosis during electrophoresis, and also cause considerable undesirable adsorption effects during gel chromatography. Despite these disadvantages, the gels have a large porosity, and have been used to separate virus particles according to size.

The use of granular agar is preferred, and two procedures have been described for the preparation of agar and agarose beads.

First: solutions of agar are suspended at 50°C in the form of spheres in a benzene/toluene mixture by the use of stabilizing agents and proper stirring. The spheres will gelatinize on cooling and can be isolated in the swollen state. The concentration of the carbohydrate can be varied between 1 and 15 per cent to give the required material (Hjerten, 1964).

Second: agar, molten by autoclaving for 20min. at 120°C and cooled to 65°C, is forced through a narrow glass jet into ice-cold water by a pressure of 1 to 2 atm. Ether is subsequently layered above the ice-cold water, and droplets of agar form a gel as soon as they come into contact with the cold ether. They sink through the ether and can be collected periodically from the water phase (Bengtsson and Philipson, 1964). The removal of agaropectin containing the ionic groups from the agar yields agarose, which is a far better matrix for the gel. However, the isolation process is not easy. Cetylpyridinium chloride was used to precipitate the acidic components (Hjerten, 1962), but this agent was difficult to remove subsequently. The two components (agaropectin and agarose) can also be separated by selective precipitation from glycol (Russell, Mead and Polson, 1964). Neither of the two methods gave an agarose which was completely sulphur free. Agarose beads can be obtained by the same procedures as outlined for agar, but are now available commercially in the swollen state. This is because, in contrast to the Sephadex gels, these substances should not be dried. Agarose gels, due to their large pore size (Sepharose 2B has an approximate fractionation range of 8×10^4 to 20×10^6 as measured by the fractionation of dextrans) can be used to separate large molecules, and have been used for nucleic acids (see p.130). However, because of the high elasticity of the beads, a gel bed will compress at high flow rates. Furthermore, the gel systems are crosslinked by hydrogen bonds so that any agent which is a potential hydrogen-bond breaker should not be included in the eluting medium (e.g. urea). Organic solvents should not be used and very strong salt solutions should be tested with a small quantity of the agarose in order to determine their compatibility. Agarose, being a natural product, is susceptible to bacterial degradation, and the eluting medium should always contain 0·02 per cent sodium azide to prevent bacterial growth.

Column parameters

A packed column of gel has certain parameters which define the limits of fractionation. The total volume of the gel bed (V_t) consists of the volume of the gel matrix (V_m), the volume between the gel granules (V_o = outer volume) and the volume contained within the gel granules (V_i = internal volume):

$$V_t = V_m + V_o + V_i$$

The individual parameters can be determined, and approximate values have been given by the manufacturers, for each gel type, e.g. 'Sephadex G-200' (for 1g. Sephadex after swelling in water.), V_t = 30ml., V_o = 9 ml. and V_i = 20ml. However, for more accurate work these must be estimated experimentally for a particular column used.

The total volume (V_t) can easily be determined by measuring the volume of the column from the base to the gel surface, using the internal diameter of the column and length of the gel bed, and the outer volume (V_o), can be ascertained by passing through the column a substance of sufficiently high molecular weight so that it cannot penetrate the internal volume. Thus the volume of solvent collected from the column between the commencement of elution and the emergence of this substance, corresponds to the volume of liquid between the gel granules, and is known as the void volume.

The internal volume (V_i) for the types of gel so far discussed may be calculated from the dry weight of the gel (X), the solvent regain (S) and the density of the solvent (P) thus:

$$V_i = X \times S/P$$

The matrix volume (V_m) may then be calculated from the first equation, or neglected completely in the case of strongly swollen gels. There is one other important parameter: the elution volume (V_e) of a substance. This is defined as the amount of solvent collected from the time of sample application to that when the position of a particular substance is at its maximum concentration. This value can only usually be obtained by direct measurement, and is strongly dependent on the particular column used. It is useless therefore to try to compare V_e values obtained on one column with values obtained on another. On the other hand, the ratios V_e/V_t and V_e/V_o are both independent of the column geometry, but are sensitive to differences in packing density. Occasionally the value for V_i, obtained by subtracting the V_e values of a very large and a very small molecule, is significantly smaller than the theoretical value obtained as described previously. The traditional explanation for this is that a certain amount of solvent is required to solvate the polymer chains, and that this solvent is not available for diffusion. Thus only a fraction (K_d) of the internal volume is available to small molecules, and for these molecules

$$V_e = V_o + K_d . V_i$$

K_d is constant for a given gel, and is thus independent of the column used and may be calculated from $K_d = V_e - V_o/V_i$ (if V_i is known from the previous calculation).

Gel filtration of nucleic acids

The nucleic acids, as in the case of proteins, have a large range of molecular weights. These range from low molecular weight tRNA molecules to very large DNA chains from bacterial and mammalian chromosomes. Thus the porosity of any one gel cannot satisfactorily discriminate between all nucleic acid types. The lower porosity gels, such as 'Sephadex G-25', were first used in the desalting of DNA solutions, and also in the removal of phenol after DNA isolation (Shepherd and Petersen, 1962). RNA and phenol could also be separated in a similar way as shown by Gross, Skoczylas and Turski (1965). DNA, because it is completely excluded from these smaller pore size gels, could be separated from tRNA by filtration through 'Sephadex G-200' (Bartoli and Rossi, 1967) and, provided the sample volume was less than 10 per cent of the internal volume of the column gel, then up to 15mg. of DNA (I) and tRNA (II) could be separated on Sephadex G-200 in 0·1 M sodium acetate with 92 to 96 per cent recovery (Ahonen and Kulonen, 1966) (Figure 5.1). No separation could be observed if Sephadex G-75 was used.

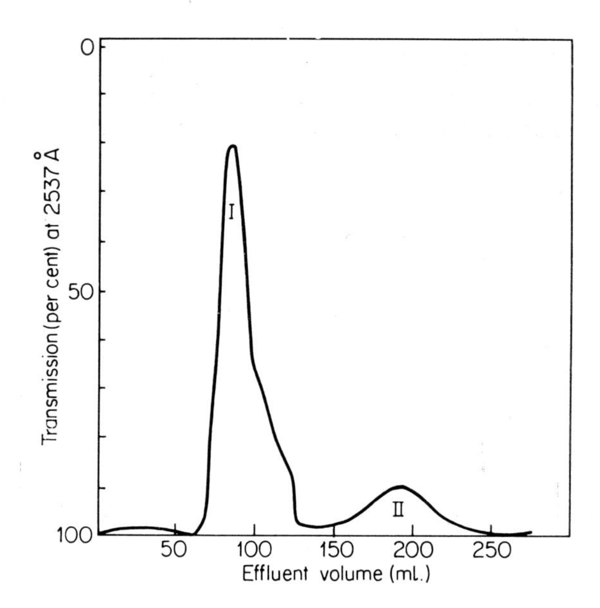

Figure 5.1. Separation of DNA (fraction I) and tRNA (fraction II) on Sephadex G-200 column. The details and the identification of the peaks are described in the text. (After Ahonen and Kulonen (1966))

Sephadex G-100 eluted with a decreasing gradient of NaCl will partially resolve tRNA from *E. coli* into two main peaks of activity. However, in the absence of salt, or in M NaCl, the fractionation was very poor (Roschenthaler and Fromagenot, 1965). In contrast, Schleich and Goldstein (1964) employed Sephadex G-100, equilibrated with M NaCl, to resolve both crude and countercurrent-distributed tRNA from *E. coli* into several molecular species (Figure 5.2). Only peak 3 was found to have any amino acid-acceptor activity, but the inactive species could be converted to active material by urea treatment. These authors suggested that aggregation of the tRNA molecules was taking place.

In a later paper (Schleich and Goldstein, 1966), a phenol preparation of tRNA (eluted from DEAE-cellulose) was fractionated on G-100 using M NaCl as the eluting medium in a similar manner as described above (Figure 5.3). The elution profile showed four peaks, the major one (peak 4) being tRNA, which consisted of 75 per cent of the total material. The

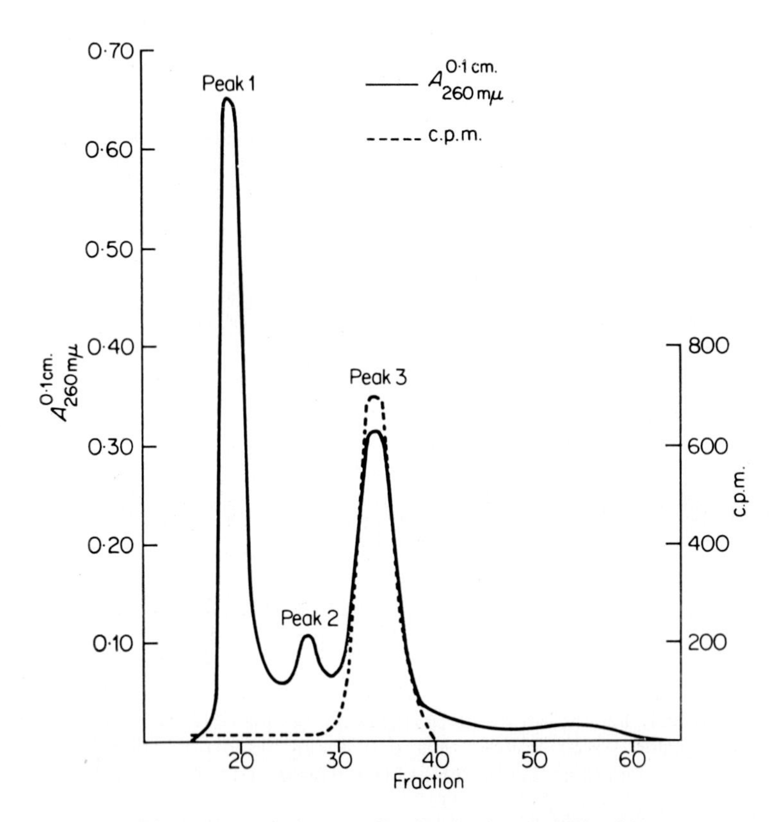

Figure 5.2. Elution profile (Sephadex G-100) of low partition coefficient tRNA highly specific for alanine and proline acceptance obtained from a 1889 transfer CCD. Counts per minute (c.p.m.) is the expression of alanine acceptor activity per aliquot of 0·15 ml. column effluent. Proline acceptor activity (not shown) tracks its alanine counterpart. (After Schleich and Goldstein (1964))

other nucleic acids were identified as ribosomal RNA (peak 1), mRNA (peak 2) and an RNA (peak 3) which, on the basis of nucleotide composition, end groups and chain-length, appears to be identical to the 5S RNA of ribosomes.

Polynucleotides have been separated and their chain-length determined by gel filtration on Sephadex G-25, G-50 and G-75 (Hohn and Schaller, 1967). In the case of the higher polynucleotides, separation was determined by chain-length. However, in the case of short-chain polymers, base-specific interactions and the presence or absence of a terminal phosphate, influenced the eluting behaviour.

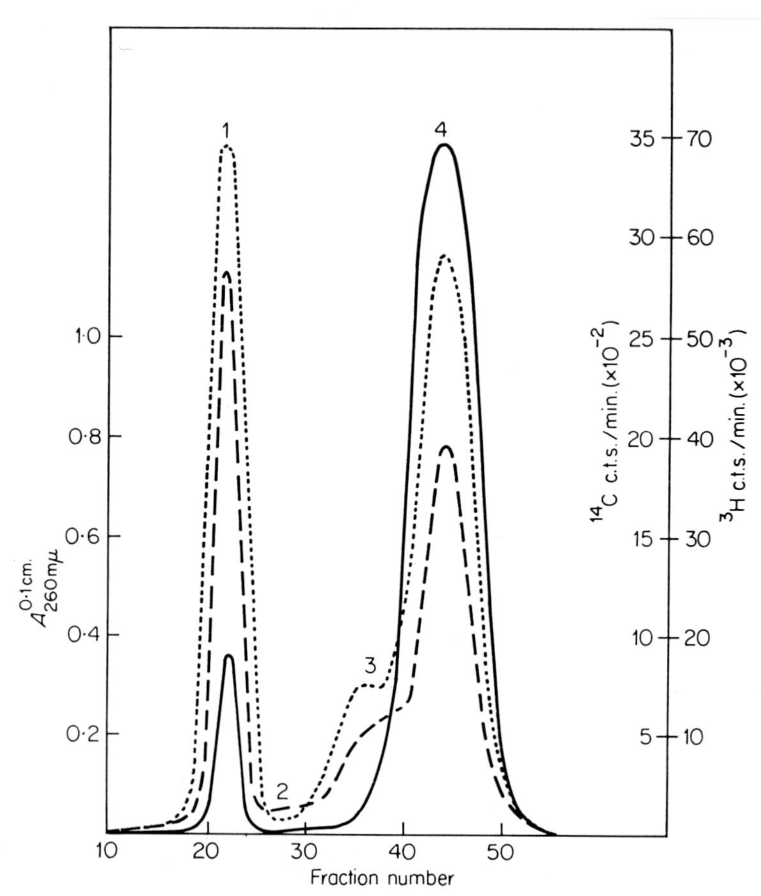

Figure 5.3. Gel filtration effluent profile (Sephadex G100) of *in vivo* labelled (^{14}C)methyl (^{3}H)uracil tRNA from *E. coli* K12 W6. -- Absorbancy at 260 mμ; —, ^{14}C c.t.s./min; ... ^{3}H c.t.s./min. (After Schleich and Goldstein (1966))

The use of gel filtration to separate molecules with molecular weights above 2×10^5 was made possible when granulated agar gels were introduced by Polson (1961). These gels had certain disadvantages, as already discussed, including low flow rates with agar concentrations of less than 3 per cent. Granulated agarose gels of 1·5 per cent, however, were able to separate an artificial mixture of T_2 DNA and *E. coli* RNA (Boman and Hjerten, 1962), and the introduction of pearl-condensed (spherical) agarose by Hjerten (1964) further improved the fractionation using these large-pore gels.

Gel filtration of poliovirus RNA on 2 per cent pearl-condensed agarose was dependent on the ionic composition of the eluting buffers (Oberg, Bengtsson and Philipson, 1965). Lithium phosphate (10^{-3} M) eluted the RNA, which had a molecular weight of 2×10^6, at the void volume of the column, while 10^{-2} M lithium phosphate containing 10^{-3} M $MgSO_4$ eluted the RNA at 60 per cent of the bed volume (Figure 5.4). In the case

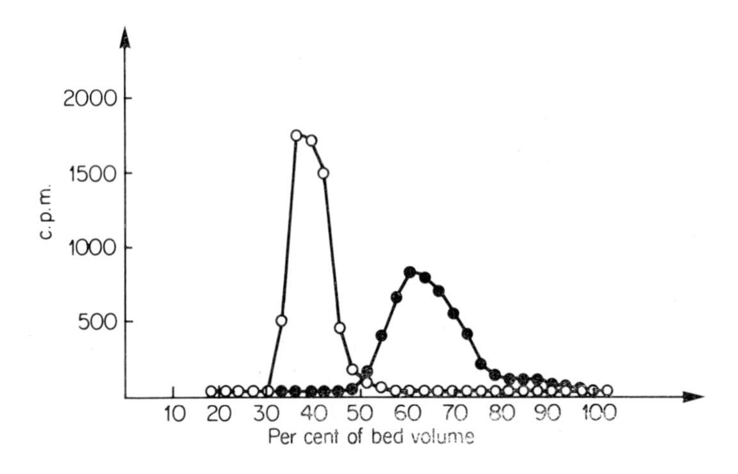

Figure 5.4. Gel filtration of poliovirus RNA labelled with ^{32}P on a 2 per cent agarose column, 1·8 × 26 cm., at room temperature in different buffers. The flow rate was 2 ml./cm^2./h. and 2 ml. fractions were collected. The buffers were 10^{-3}M lithium phosphate, pH 6·0 (open circles) and 2×10^{-2}M lithium phosphate pH 6·0 with 2×10^{-3} M $MgSO_4$ (filled circles). (After Oberg, Bengtsson and Philipson (1965))

of 4 per cent gels, the DNA and RNA eluted together at the void volume of the column, but tRNA was retarded and eluted much later. The sieving effect can be used to separate and purify replicative (RNase-resistant) RNA intermediates of R17 bacteriophage from cellular RNA of

E. coli. Using 4 per cent agarose beads the ^{3}H-labelled RNA eluted at the void volume of the column, whereas the other RNA's eluted more slowly (Erickson and Gordon, 1966).

It is therefore more difficult to use individual gel types of differing exclusion limits (i.e. range of molecules fractionated according to the limits imposed by the pore size) for nucleotides and nucleic acids.

Recently Agarwal and coworkers (1970), in Khorana's laboratory, have managed to synthesize the complete gene for alanine-tRNA from yeast. This remarkable achievement exploited the natural ability of polynucleotides to align themselves by using the base-pairing mechanism. The 'gaps' were then joined using polynucleotide kinase and ligase, and in this way, chemically synthesized segments of DNA oligomers were combined into a piece of native double-stranded DNA, which corresponded to the gene alanyl-tRNA. During the course of his work, it was often necessary to separate the joined products from the starting materials, and the method they chose was to subject the reaction mixture to gel filtration on agarose columns. This separated the desired newly formed compound, which had a greater molecular size, from the lower molecular weight precursors.

Polyacrylamide gel electrophoresis

Introduction

The development of synthetic polymers such as the acrylamides as support media for electrophoresis has added greater sensitivity and resolution to the earlier techniques of paper and starch gel electrophoresis. Gels with differing pore size can be prepared which act as molecular sieves. The various components in a particular sample are first concentrated into thin starting zones, and subsequently separated according to (*a*) charge and (*b*) size and shape, by a combination of electrophoresis and molecular sieving.

The polyacrylamide gel columns originally used consisted of three layers: (1) a large-pore (sample) gel containing the sample ions in which the electrophoretic concentration of these ions was initiated; (2) a large-pore (spacer) gel in which the electrophoretic concentration of the sample ions was completed; and (3) a small-pore (running) gel in which the electrophoretic separation took place. The pH values of the running gel, spacer gel and the electrophoresis buffer were different (discontinuous system) and critical for an effective separation. The theory regarding the choice of pH is extremely complex (Ornstein, 1964), although a simple strategy for the design of a discontinuous electrophoretic system was provided by Williams and Reisfield (1964).

Various modifications of the original system are now used. In particular, it has been found that: (*a*) the sample and spacer gels can be omitted and the biological sample (dissolved in sucrose to prevent backward diffusion) applied directly to the running gel, and (*b*) the use of ethylene diamine tetraacetate (EDTA), which has a high electrophoretic mobility, precludes the use of a discontinuous buffer system. The modifications have been used by Loening to fractionate nucleic acid species, and an outline of his method (Loening, 1967, 1968a) is given below.

Materials and methods

Purification of monomers: acrylamide (70g.) is dissolved in analar chloroform (1l.) at 50°C and the solution filtered hot without suction. The crystals separate on slowly cooling the filtrate to —20°C, and can be recovered by filtration in a chilled funnel. They are then washed with cold chloroform and dried.

Bis-acrylamide (10g.) is similarly recrystallized from acetone (1l.) and the pure crystals washed with cold acetone before drying. These compounds are extremely poisonous and care must be taken against contact with the skin or inhalation of the light crystals.

Stock solutions containing 15 per cent (w/v) acrylamide and suitable concentrations of bis-acrylamide in water, are stable for at least one month if stored at 5°C in the dark.

Recrystallization (*a*) prevents RNA from sticking to the gel on electrophoresis and (*b*) removes u.v. absorbing materials which would interfere with the subsequent scanning of the resolved nucleic acids at 260 to 265 mμ.

Preparation of the gels. The sieving effect of a gel depends on the concentration of acrylamide, and on the degree of crosslinking with *N,N′*-methylene bis-acrylamide. The choice of a suitable monomer to 'bis' ratio, however, is limited by the need to obtain a 'manageable' and stable gel. The bis-acrylamide concentration is usually 5 per cent of that of acrylamide for a 2 to 5 per cent (w/v) gel and is 2·5 per cent that of acrylamide for a 5 to 8 per cent (w/v) gel.

Running gel solutions are prepared from 15 per cent stock solutions by suitable dilution with water and the appropriate buffer. The acrylamide/water/buffer mixture is degassed at room temperature in vacuo for 15 sec. *N,N,N′,N′*-tetramethyl ethylene diamine (TEMED), an initiator of polymerization, and ammonium persulphate, a polymerization catalyst, are added and the resulting solution immediately pipetted into vertical gel tubes (¼in. diameter by 2½ or 5in. long). The tubes are usually made of perspex since the gels tend to adhere to a glass surface, but glass is adequate for the more concentrated (4 to 8 per cent) gels. Water is then carefully layered on the gel surface to ensure a flat gel surface, and the gel allowed to polymerize at room temperature, the time of polymerization increasing with decreasing gel concentration. Rubber rings are usually inserted into the bases of the tubes after polymerization to prevent the soft gels from sliding out during electrophoresis. The running buffer normally employed for RNA separation contains 36 mM tris, 30 mM NaH_2PO_4 and 1 mM EDTA (disodium salt), pH 7·7 to 7·8.

Electrophoresis. Electrophoresis is carried out either at 0 to 5°C using the pH 7·8 buffer (the same concentration of ions as in the running gel), or at room temperature using the pH 7·8 buffer containing 0·2 per cent sodium dodecyl sulphate (SDS) to inhibit any nuclease activity. The vertical equipment described by Davis (1964) is usually used. The gels are prerun for ½ to 1h. at 5 mA. per tube partly to remove u.v. absorbing material and polymerization catalysts, and to allow SDS to enter the gel. The nucleic acid sample (50–100 μg.), dissolved in the pH 7·8 buffer (10–100 μl.) containing 5–6 per cent sucrose, is then layered over the gels, and the electrophoresis continued at 5 mA per gel for a time dependent on the resolution required and the length of gel.

After completion of the electrophoretic run, the rubber rings are removed, and the gels gently blown out of the tubes with a rubber teat. The dilute gels cannot be handled without damage, and are therefore usually picked up by sucking into tubes of the same diameter.

Scanning of ultraviolet-absorption. The gels prepared from purified solutions have an absorption of less than 0·3 at 265 mμ. They can be scanned in quartz cells in either a Chromoscan or U.V. Polyfrac (Joyce-Loebl) instrument, using a medium-pressure mercury lamp and special filters (Loening, 1969; see also a reprint on the fractionation of RNA on polyacrylamide gels and scanning of gels in the u.v. region, published by Joyce-Loebl).

Scanning of radioactivity. Indian ink lines (one near each end of the gel) are injected with a fine syringe at right angles to the gel length to act as a marker for the registration of the optical and radioactivity scans. After optical scanning, the gels are frozen in an open aluminium trough in dry ice, and sliced into thin sections (0·25mm. to 1mm.) which are then dried either on filter paper (in the case of ^{32}P or ^{14}C), or by heating in 10 per cent (v/v) piperidine at 60°C (for ^{3}H), before the addition of scintillator for radioactive counting.

Biological applications

Fractionation of low molecular weight RNA

Gels with a concentration of 5 per cent or higher have been used to separate the degradation products produced by ribonuclease action on high molecular weight ribosomal RNA.

The specific cleavage of yeast ribosomal RNA with ribonuclease was studied by McPhie, Hounsell and Gratzer (1966) and the course of digestion is shown in Figure 5.5. They found a linear relationship between the mobilities of the RNA components on a 5 per cent gel and their sedimentation coefficients. They also concluded that labile points (hot spots) existed in the polynucleotide chain, and that the number of such points could be deduced from the number of bands generated:

$$Z = \tfrac{1}{2}\,(n + 2)\,(n + 1)$$

where Z is the number of components including the undegraded species and n is the number of hot spots.

Gould (1966) made a similar study of the ribonuclease digestion of ribosomal RNA isolated from each of the two sub-units of reticulocyte

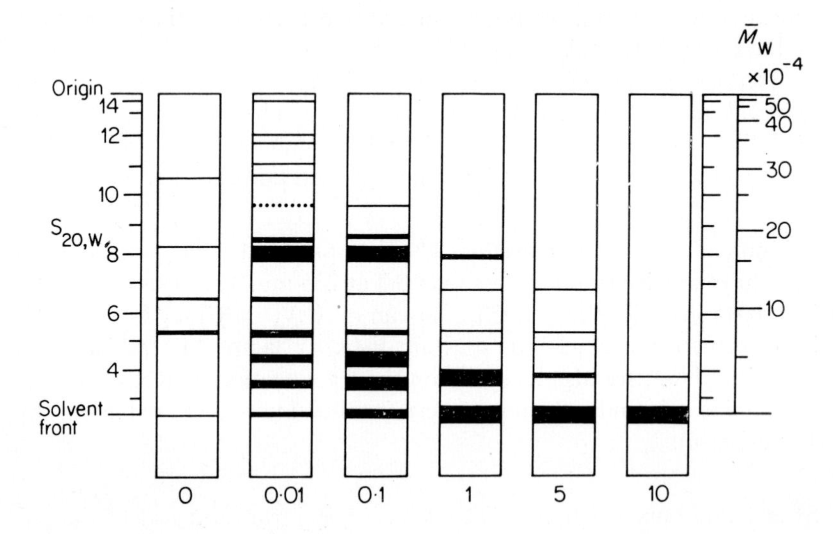

Figure 5.5. The course of Tl-ribonuclease digestion of yeast ribosomal RNA fractionated by polyacrylamide gel electrophoresis. The figures are the amounts in milligrams of enzyme added to 0.5 mg. of RNA samples all of the same concentration (10 mg./ml. final). The scale shows the sedimentation coefficients corresponding to R_F values. The molecular weight scale is based on the relation between $S_{20,w}$ and $\bar{M}_w$. (After McPhie, Hounsell and Gratzer (1966) *Biochem.* **5**, 998. Copyright by the American Chemical Society. Reprinted by permission of the copyright holder)

ribosomes. The patterns of degradation were different for the two types of RNA, implying that the regions susceptible to attack were distributed at characteristic intervals in each case. The initial digestion products were relatively large with molecular weights in the 250,000 to 400,000 range. As digestion proceeded, the high molecular weight species were degraded into smaller fragments in the 20,000 to 250,000 range. Using the equation of McPhie and coworkers (1966), Gould deduced that there were approximately five vulnerable regions for the 30S ribosomal RNA, and three for the 19S component. It was suggested that these sites were preferentially attacked because the RNA chain is folded in a way which leaves them exposed specifically.

Richards, Coll and Gratzer (1965) also obtained a linear relationship between the mobilities of the 4S amino acid-tRNA and its degradation products and their sedimentation coefficients. They showed that gel electrophoresis could be used to purify the tRNA from any degraded material (Figure 5.6), the major β-band being associated with the amino acid-accepting activity of tRNA.

Fractionation of high molecular weight RNA

General considerations: investigations of the high molecular weight ribosomal RNA species by gel electrophoresis have been made mainly by Loening (1967), using the method outlined above. However, experiments carried out simultaneously by Grossbach and Weinstein (1968) using a discontinuous buffer system, and a specific staining procedure gave similar results. Several conclusions were reached from these early experiments:

(*a*) The mobilities of the various resolved components were inversely proportional to their sedimentation coefficients (as in the case of small molecular weight RNA).

(*b*) The 28S and 18S species from higher organisms, and the 23S and 16S species from bacteria, were best resolved on 2·4 to 2·5 per cent gels.

(*c*) The separation between the small (16S or 18S) ribosomal RNA and the 4S and 5S species increased as the gel concentration increased above 2·4 per cent.

(*d*) 4S and 5S could be resolved on a 7·5 per cent gel.

(*e*) The differences in the charge to mass ratio of the various RNA species are probably too small to affect the mobilities, and hence separation occurs mainly by a sieving effect.

(*f*) Changes in secondary structure should affect mobility: an increase in volume due to chain unfolding should increase the resistance to the passage of the molecule through the 'sieve', and hence decrease its mobility. As a direct corollary of this, differences in base composition, which dictate the resistance of the polynucleotide chain to unfolding, should be detected under certain experimental conditions. The results

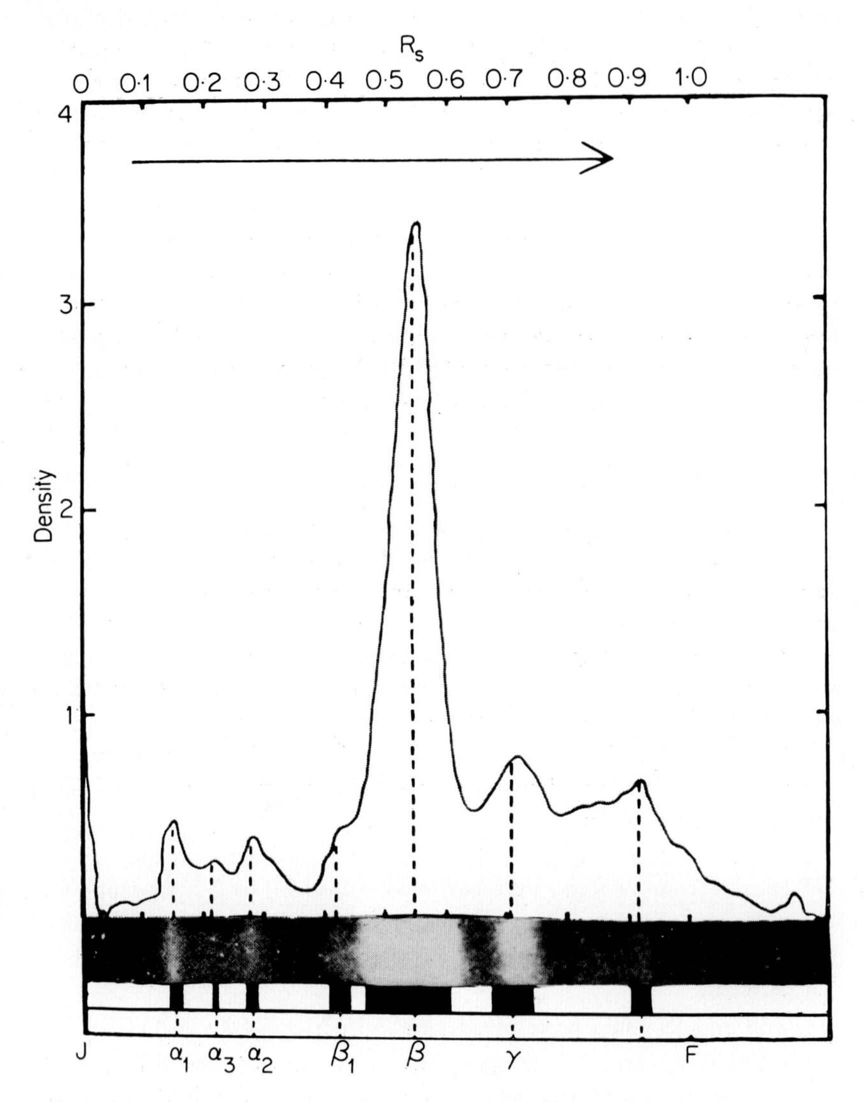

Figure 5.6. Photograph and densitometer trace of 10 per cent gel with 200 μg. tRNA. J indicates junction of small and large pore gel and F the anion front. (After Richards, Coll and Gratzer (1965) *Anal. Biochem.* **12,** 452. Copyright by the Academic Press Inc.)

obtained by Loening (1967) illustrate points (*b*) to (*d*) described above (Figure 5.7). These points and the remaining conclusions will be further examined below.

Recently Jaenisch, Jacob and Hofschneider (1970) have demonstrated the synthesis of a specific mRNA by bacteriophage M13 infected *E. coli*. Figure 5.8 shows the fractionation of rapidly-labelled RNA from M13 infected *E. coli* on polyacrylamide gel. Four to five peaks were observed which hybridized with heat-denatured M13 replicative form of DNA (Figure 5.8). The authors were able to estimate the half-life of M13 mRNA as approximately 18 min.

Acrylamide gel electrophoresis has proved extremely useful in testing hypotheses on evolution. Green leaf tissues possess two distinct ribosome species: an 80S species (similar in size to animal ribosomes) which is present in the plant cytoplasm, and a 70S species (similar to bacterial ribosomes) which is restricted to the chloroplast. This suggests the presence of more than the usual two ribosomal RNA species (found in non-green tissues), and indeed the total RNA from green tissues could be fractionated into three to four high molecular weight and four small molecular weight components by MAK chromatography (Dyer, 1967). Loening and Ingle (1967) made a more intensive investigation of this problem, and obtained good resolution of the various species. Figure 5.9, illustrates the location of the various types of ribosomal RNA found in green tissues. Two species only, 25S and 18S, were observed in the root and hypocotyl tissue, but two additional major components (23S and 16S), and one minor (13S) component, were observed in the cotyledons. Studies on a variety of plants showed the 23S and 16S species to be characteristic of all green but not non-green tissues. Moreover, the 23S and 16S rRNA species increased relative to the 25S and 18S during development and maturation of the leaf, and some additional minor components (21S, 15S and 13S) appeared. These changes did not occur if the plants were grown in the dark, suggesting they were associated with the chloroplast, and gel electrophoresis of chloroplast RNA did in fact show an enrichment of the 23S and 16S components (Figure 5.10).

A comparison of the ribisomal RNA species from green tissues with those from *E. coli*, pea root tip and blue-green algae (Figure 5.11) indicated that the RNA of blue-green algae was similar to that of *E. coli* and chloroplasts. This supported the hypothesis that chloroplasts have evolved from symbiotic blue-green algae (Loening and Ingle, 1967).

The presence of a 25S species in plant cytoplasm, as opposed to a 28S fraction in mammalian cell cytoplasm, indicated that three types of ribosomes exist:

(*a*) 80S in mammalian cells containing 28S and 18S RNA.

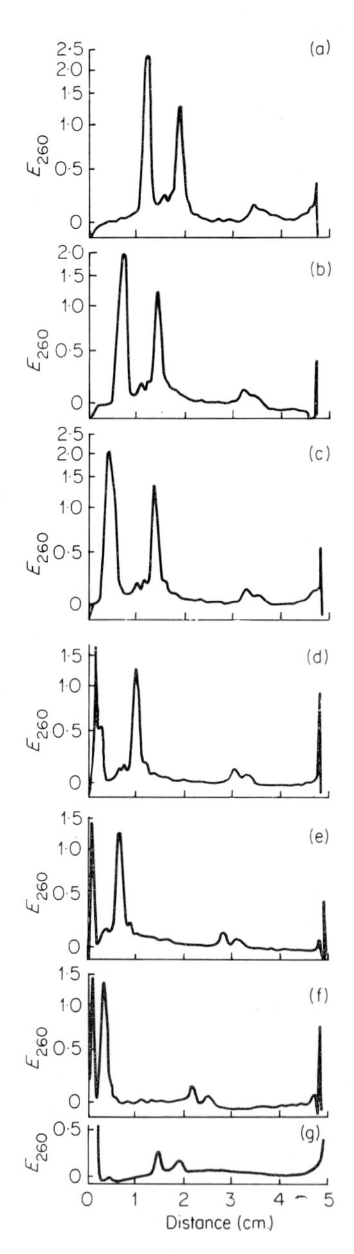

Figure 5.7. Electrophoresis of RNA at different gel concentrations. RNA was prepared from rabbit reticulocyte polysomes by the sodium triisopropylnaphthalenesulphonate-phenol-*m*-cresol procedure. In each case 40 μg. of RNA was applied/gel; gel concentrations were: (a) 2·2 per cent; (b) 2·4 per cent; (c) 2·6 per cent; (d) 3·0 per cent; (e) 3·5 per cent; (f) 5·0 per cent; (g) 7·5 per cent. Electrophoresis was carried out for 65 min. at 50 v. in buffer. (After Loening (1967))

(*b*) 80S in plant cytoplasm containing 25S and 18S RNA.

(*c*) 70S in bacteria and plant chloroplasts containing 23S and 16S RNA.

The size of the rRNA of plants is thus intermediate between those of bacteria and mammals, and it appears that the size within the different kingdoms has been conserved during evolution. This apparent division into distinct classes was tested by Loening (1968b) using rRNA from a wide range of species. In particular, species were selected which would be expected to show intermediate sizes of rRNA on biological grounds. A

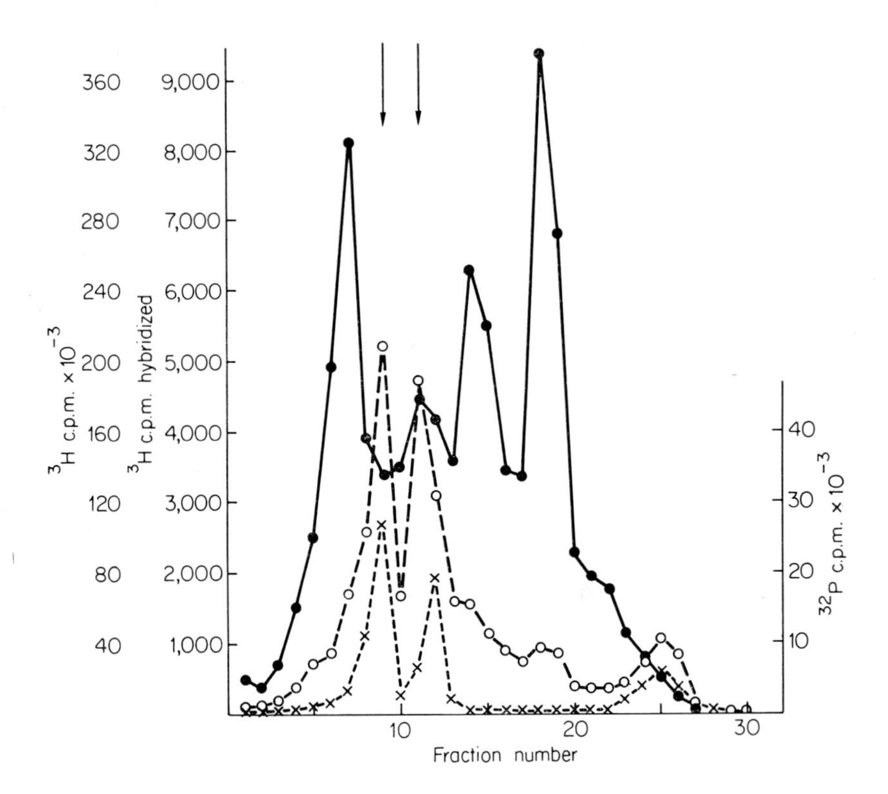

Figure 5.8. Gel electrophoresis of RNA from M13 infected *E. coli* AB 301cells with subsequent hybridization of the eluate of the single gel fractions with M13 RF DNA. o– – –o, Total TCA precipitable ^{3}H radioactivity from a 3 min. ^{3}H-uracil pulse; x– – – – – –x, ^{32}P radioactivity originating from uninfected cells and used as an internal standard; ●—●, ^{3}H radioactivity hybridized to denatured M13 RF II DNA. The arrows indicate the positions of the 23S and the 16S ribosomal RNA. Migration was from left to right. (After Jaenisch, Jacob and Hofschneider (1970))

direct linear relationship was observed between the mobilities of the various rRNA species and the logarithm of the molecular weight (mobility $\propto 1/\log$ molecular weight). The results obtained using this relationship and the observed electrophoretic mobilities of various species showed that, with very few exceptions:

(*a*) Bacteria, actinomycetes, blue-green algae and higher plant chloroplasts all had the 23S and 16S species with molecular weights of $1{\cdot}1 \times 10^6$ and $0{\cdot}56 \times 10^6$, respectively.

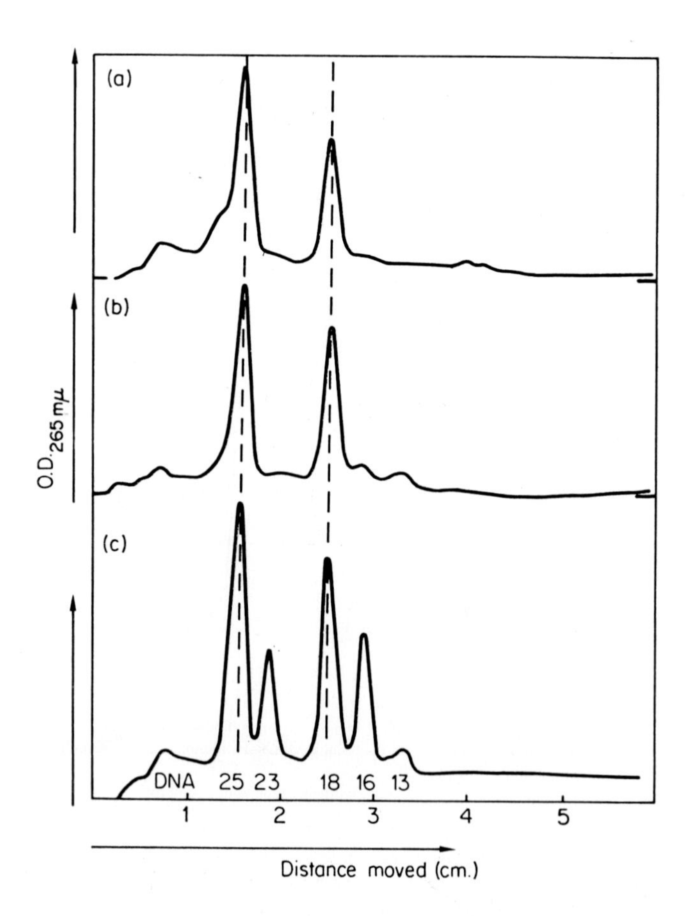

Figure 5.9. Gel fractionation of nucleic acids from radish. Fifty radish seedlings grown in the light for 6 days, were separated into roots (a), hypocotyls (b) and cotyledons (c) and total nucleic acids were prepared. (After Loening and Ingle (1967))

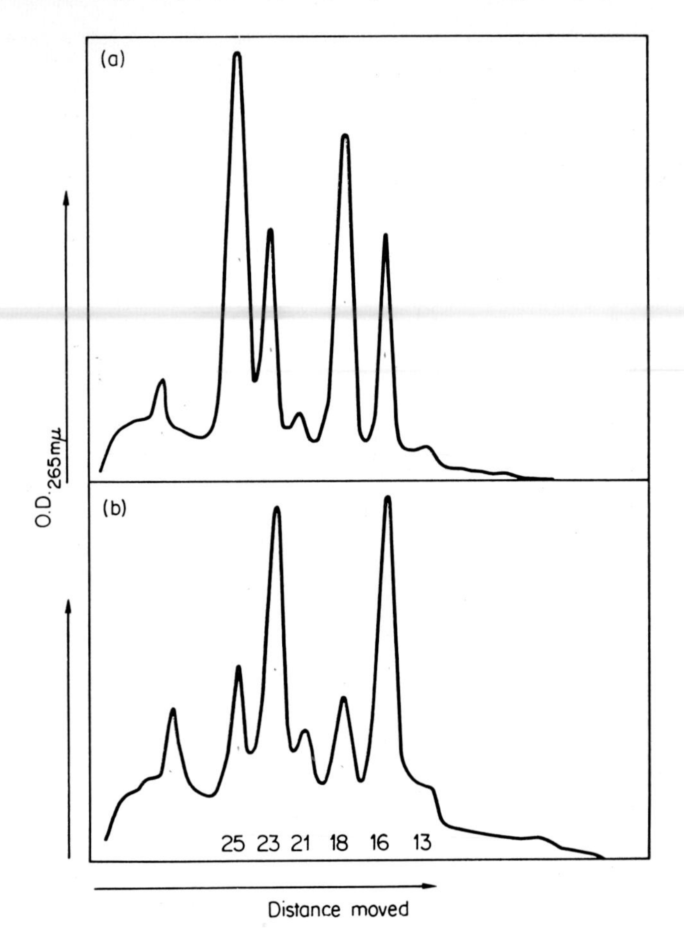

Figure 5.10. Chloroplast nucleic acids. Nucleic acids were prepared from the whole leaf (a) and from the chloroplast fraction (b). Leaves from 8-day-old French bean seedlings were ground by pestle and mortar at 0°C in a medium containing 0·4 M mannitol, 50 mM tris, 50 mM potassium chloride, 5 mM magnesium acetate, 0·1 per cent bovine serum albumin and 0·1 mg./ml. of Cleland reagent pH 7.8 at 0°C. The homogenate was filtered through cheese cloth and centrifuged at 100 g. for 4 min. The supernatant was centrifuged at 1,000g. for 5 min. to pellet the chloroplasts. The chloroplasts were washed by resuspending in the medium and collected by centrifugation at 1,000g. for 5 min. The pellet was suspended in a medium containing 1 per cent triisopropylnaphthalene sulphonate, 6 per cent *p*-aminosalicylate and 1 per cent sodium chloride, and nucleic acid was extracted with phenol. (After Loening and Ingle (1967))

(*b*) The higher plants, ferns, algae, fungi and some protozoa possessed a 25S and 18S rRNA with molecular weights of $1{\cdot}3 \times 10^6$ and $0{\cdot}7 \times 10^6$.

(*c*) All animals possessed the smaller $0{\cdot}7 \times 10^6$ rRNA common to higher plants, but the large rRNA component was '28S'.

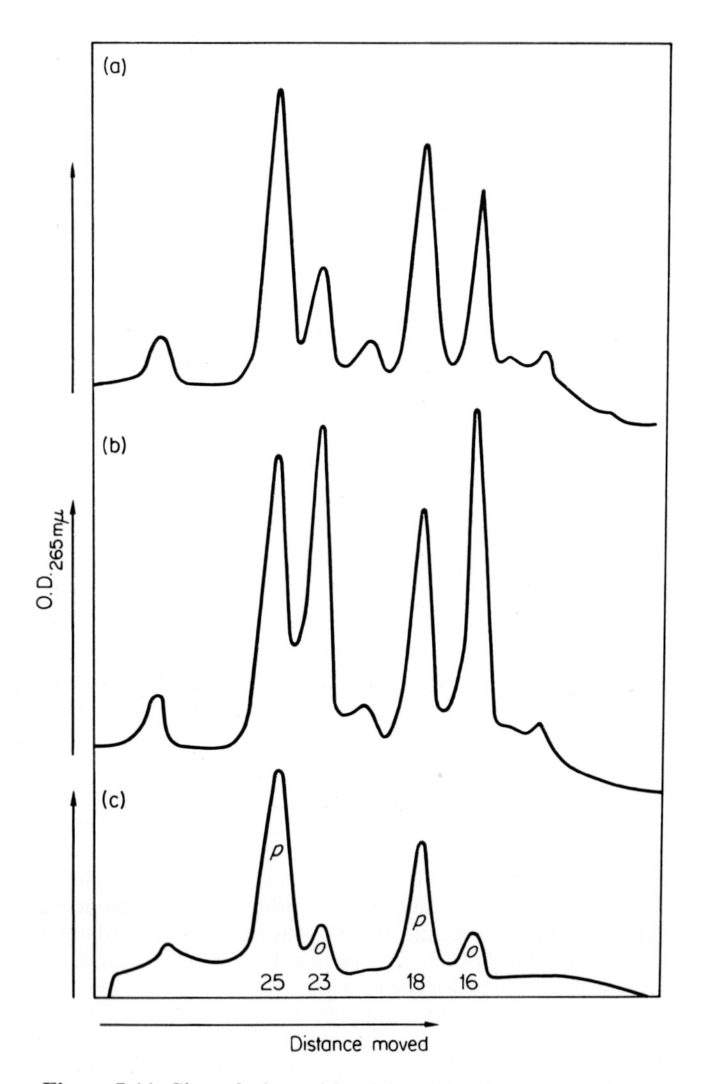

Figure 5.11. Size of plant chloroplast RNA's compared with bacterial and blue-green algal RNA's. (a) Nucleic acids from 8-day-old French bean leaf; (b) French bean leaf plus *E. coli* RNA's; (c) pea root tip (*p*) plus a species of *Oscillatoria* (*o*) RNA's. (After Loening and Ingle (1967))

(*d*) The '28S' species increased from a molecular weight of $1{\cdot}4 \times 10^6$ in sea urchins to $1{\cdot}75 \times 10^6$ in mammals.

(*e*) Two exceptions out of the many species examined are *Euglena* and *Amoeba*: these possess exceptionally high molecular weight RNA which is also unstable.

The determination of the molecular weight of the various RNA species on gels by the relationship: electrophoretic mobility $\propto 1/\log$ (molecular weight), is possible only because differences in the charge to mass ratio of the various species are small, and hence the sieving effect of the gels is the primary factor governing migration. However, to interpret the significance of the observed RNA separations, one must examine what effect (if any) the conditions of electrophoresis have on the mobilities, and hence the apparent molecular weight of the RNA species. Since the sieving effect of gels depends not only on the size of molecules (weight), but also on the effective molecular diameter, any conformational change will be manifested in a change of mobility. RNA molecules unfold at low ionic strength, the degree of unfolding being more pronounced in those molecules which possess a low content of guanine and cytosine relative to the other bases, uracil and adenine. This unfolding increases the effective diameter of the molecule, and should lead to a decrease in mobility. In theory, therefore, it should be possible to detect differences in base composition between the various RNA species of similar molecular weight. Loening (1969) tested this hypothesis by examining the mobilities of the rRNA molecules of *Drosophila* and *Xenopus* which have similar weight but quite different GC contents. A mixture of the two species was fractionated using buffers of low and medium ionic strength and in a low ionic strength buffer containing Mg^{2+} ions (to maintain the RNA in a compact form). Figures 5.12 and 5.13 show the observed separations. In the normal electrophoresis buffer (medium ionic strength), the two large and two small rRNA molecules of the two species could just be distinguished. In the Mg buffer, the two smaller rRNA (molecular weight $0{\cdot}7 \times 10^6$) migrated as one unresolved peak, but the large components of $1{\cdot}41 \times 10^6$ and $1{\cdot}52 \times 10^6$ were separated according to molecular weight as in the normal buffer. In low ionic strength buffer, the rRNA components of *Drosophila*, having a lower GC content, unfolded more than those from *Xenopus* and therefore had decreased mobilities. Hence the two smaller components were readily resolved, whereas the decreased mobility of the large component of *Drosophila* caused it to migrate almost with that of *Xenopus*. Since the mobility of the small rRNA of *Xenopus* was approximately the same in the low salt and Mg buffers, and the mobilities of the large rRNA of *Xenopus* and *Drosophila* were only slightly greater in the Mg buffer than in the low salt buffer, it was concluded that the structural changes which

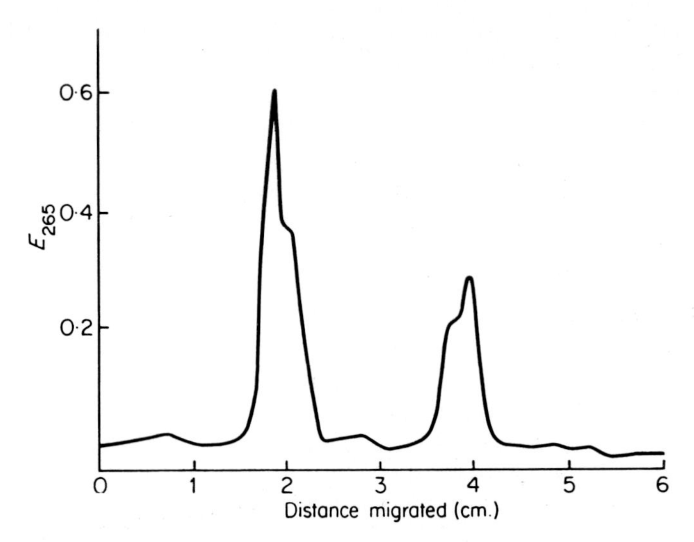

Figure 5.12. Electrophoresis of a mixture of similar rRNA species. Sample (10 μl.) of *Xenopus* rRNA (2mg./ml.) and *Drosophila* rRNA (1 mg./ml.) was applied to 2·4 per cent gel and electrophoresis was carried out for 3 h. at 24°C. The gel was scanned by a Chromoscan at 265 mμ. (After Loening (1969))

had occurred (which produced large increases in sedimentation coefficients) did not greatly affect the absolute mobilities. The relative decrease in mobility of the *Drosophila* RNA in low salt buffer, compared with that in the normal buffer, was correlated with only small changes in the sedimentation coefficients. The effect of Mg^{2+} ions in decreasing the effective molecular size, and hence increasing the mobility, appeared to be opposed by the effect of the Mg^{2+} ions masking the negative charges on the RNA molecules. The masking effect was less effective in the case of the larger rRNA molecules, which therefore showed a slightly increased mobility in the Mg buffer.

Gel electrophoresis in medium ionic strength buffers can therefore be used satisfactorily to measure molecular weights of RNA since at this ionic strength the molecular weight is relatively unaffected by nucleotide composition.

The relationship between the electrophoretic mobility and the sedimentation coefficient or molecular weight, which is found for tRNA and rRNA, is also applicable in the case of viral RNA. Thus, Bishop, Claybrook and Spiegelman (1967) were able to affect a good separation of various viral nucleic acids. They used a similar method to that used by Loening,

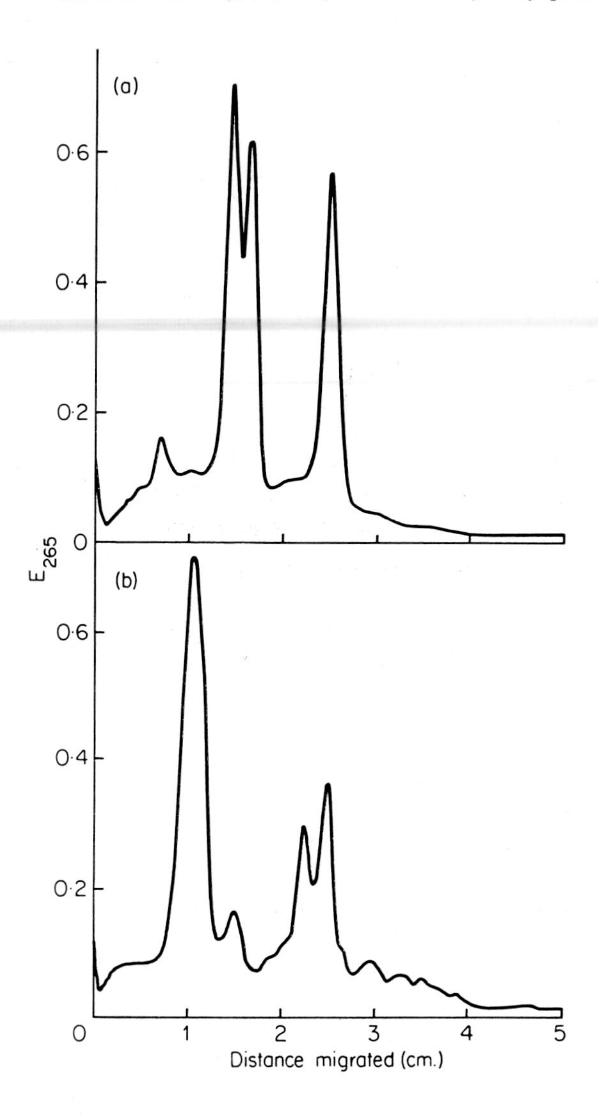

Figure 5.13. Effects of low salt and Mg buffers on the mobilities of RNA species of different base compositions. The electrophoresis of the mixture of *Xenopus* RNA and *Drosophila* RNA shown in Figure 5.12 was repeated in (a) the Mg buffer and (b) the low salt buffer. Electrophoresis was in 2·4 per cent gels for $2\frac{1}{2}$ h. at 25° (approx 3·5mA/tube at 50v.). The gels were scanned by a Chromoscan at 265 nm. (After Loening (1969))

but in addition to using gels crosslinked with bis-acrylamide, they used ethylene diacrylate crosslinked gels. Moreover, Bishop, Claybrook and Spiegelman (1967) found that if preswollen bis-acrylamide crosslinked gels were used, the mobilities of the various RNA species were increased, and the straight-line relationship was still valid. Figure 5.14 shows the

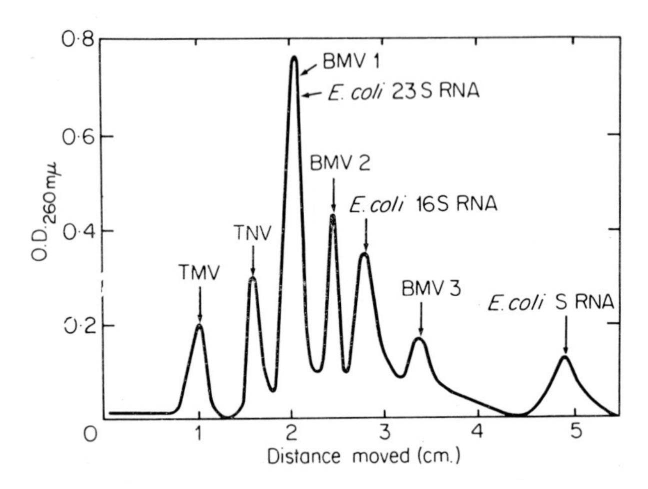

Figure 5.14. Separation of TMV, TNV, BMV and *E. coli* bulk RNA. A mixture of TMV RNA (5 μg.), TNV RNA (5 μg.), BMV RNA (10 μg.) and *E. coli* bulk RNA (10 μg.) was subjected to electrophoresis on bis-acrylamide swollen 2·4 per cent gels for 90 min. and scanned for u.v. absorption. (After Bishop, Claybrook and Spiegelman (1967))

separation of a mixture of five viral nucleic acids and *E. coli* (23S and 16S) bulk RNA. The bromograss mosaic virus (BMV) RNA was resolved into three components, each of which had a mobility, relative to satellite tobacco necrosis virus (STNV) RNA, close to that expected for the molecular weight determined by other methods. Bishop, Claybrook and Spiegelman (1967) were also able to resolve a mixture of 23S, Ø-X174 circular single-stranded DNA and 21S Ø-X174 double-stranded circular DNA (Figure 5.15).

Certain discrepancies do, however, arise when comparing some viral RNA's with rRNA, and the straight-line relationship is not always obeyed. This is particularly true of those viruses whose RNA is partially in a double-helical form due to base-pairing. The properties of this type of RNA then approach that of DNA, whose mobility appears to be almost independent of molecular weight, and varies little with gel concentration.

One important practical application of the gel fractionation of viral RNA is the retention of biological activity after electrophoresis. Qβ-RNA was purified by polyacrylamide gel electrophoresis by Bishop, Claybrook and Spiegelman (1967), and was found to be more infectious with *E. coli* spheroplasts (Figure 5.16) than was the unfractionated material, and also had a more efficient priming activity with Qβ replicase as shown by Haruna and Spiegelman (1965).

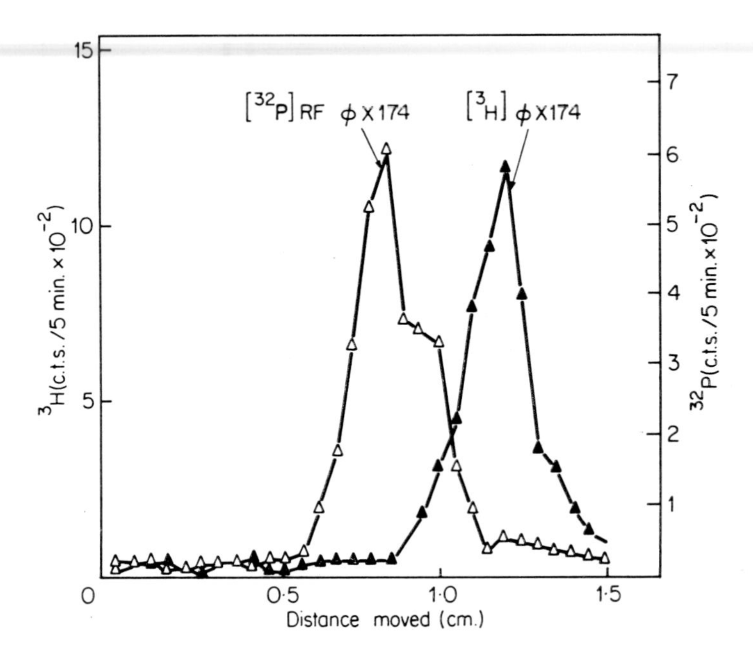

Figure 5.15. Separation of Ø-X174 single- and double-stranded DNA. Ø-X174 circular 23 S single-stranded [^{32}P] DNA (6×10^2 c.t.s./min.) was mixed with 21S Ø-X174 replicative form double-stranded, circular, [^{3}H] DNA ($2{\cdot}8 \times 10^3$ c.t.s./min.) and subjected to electrophoresis on bis-acrylamide swollen 2·4 per cent gels for 180 min. The gels were sliced and counted. (After Bishop, Claybrook and Spiegelman (1967))

Hydroxyapatite

Preparation of hydroxyapatite columns

There are several methods of preparing hydroxyapatite (HA) (Anacker and Stoy, 1958; Main, Wilkins and Cole, 1959; Jenkins, 1962 and Siegelman, Wieczorek and Turner, 1965), but the method developed by

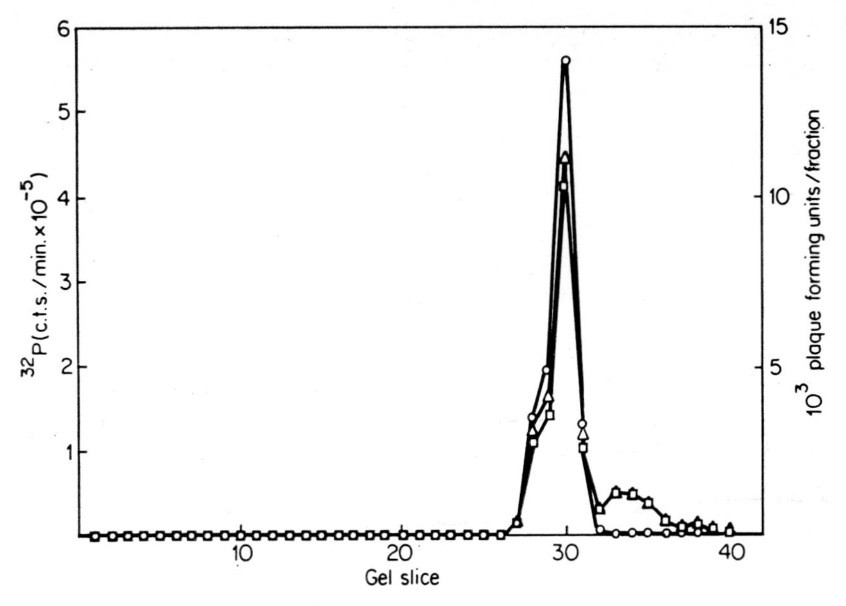

Figure 5.16. Demonstration of biological activity after electrophoresis. Qβ[^{32}P]RNA (1·1 × 10^{6} c.t.s./min.; 0·5 μg.) was subjected to electrophoresis on a bis-acrylamide (swollen 2·4 per cent) gel under standard conditions, sliced and each slice eluted twice with 0·5 ml. of 5 × 10^{-3} M-EDTA. The eluted acid-insoluble radioactivity was determined (□) and the residual acid-insoluble radioactivity in the gel slice to determine the original content of ^{32}P (△). The ability to infect spheroplasts to give plaque-forming units was determined in duplicate on portions of each eluant and the total plaque-forming units per eluant calculated (○). (After Bishop, Claybrook and Spiegelman (1967))

Tiselius and coworkers (1956) is the one usually employed, and will be cited here. Equal volumes (2 l.) of 0·5 M $CaCl_2$ and 0·5 M Na_2HPO_4, are added at a constant rate of 250ml. per h. to M NaCl (200ml.), with constant stirring. The mixture is stirred for 1h., and the resulting precipitate of calcium phosphate allowed to settle. The precipitate is washed several times with 0·005 M phosphate buffer (pH 6·7) and finally taken up into a volume of about 1 l. Sodium hydroxide solution is added (final pH 8·5), and the resulting suspension boiled vigorously for 30 min. The granular precipitate is allowed to settle rapidly and the 'fines' discarded. For high flow rates from HA columns, it is advisable to be meticulous in this discarding process. The resulting calcium phosphate crystals (hydroxyapatite) can be stored in 0·001 M phosphate buffer pH 6·8, at 4°C for several weeks.

Columns are usually packed by adding a suspension of HA in 0·001 M phosphate buffer to columns partially filled with the same buffer. When a one-centimetre layer of HA has formed, the rest of the column is packed by opening the outlet, and the HA allowed to settle under gravity flow. Elution is usually accomplished with either potassium or sodium phosphate buffer pH 6·8, using either a stepwise or linear increase in phosphate molarity up to 0·3 M. Polynucleotide samples can be loaded at widely differing concentrations ranging from 1μg. to 1mg. per ml. In most cases, the solvent is 0·001 M phosphate but 0·01 M phosphate buffer, 0·15 M NaCl, M NaCl and M KCl have also been used.

Calcium-complexing agents such as citrate or EDTA must be absent from the nucleic acid samples, but the presence of small amounts of chloroform, isoamyl alcohol, phenol, formaldehyde or urea do not seem to effect the behaviour of the columns.

Biological applications

RNA fractionation

Early attempts to fractionate RNA using HA columns were not very successful (Bernardi and Timasheff, 1961). A mixture of 20S and 29S RNA, isolated from Ascites tumour cells, was applied to a HA column and eluted with 0·2 M phosphate buffer. One component only was obtained which had an S-value of 14 indicating a decrease in molecular weight. Using stepwise elution, two peaks were obtained which eluted at 0·15 and 0·2 M phosphate. However, no differences between the RNA's of the two peaks were found. No fractionation had therefore occurred, and the separation into two peaks was an artifact due to degradation occurring during chromatography.

Pinck, Hirth and Bernardi (1968) chromatographed the double-stranded replicative form and the replicative intermediate of plant virus RNA on HA using gradient elution. Both forms were found to elute at the same phosphate molarity at which native DNA is eluted, and therefore were readily separated from plant ribosomal RNA.

Aminoacyl-tRNA from *E. coli* has also been resolved by HA chromatography by eluting with a gradient of phosphate buffer, pH 5·8 (Muench and Berg, 1966). It was found that the individual tRNA species were resolved to 80 to 90 per cent purity, and that the acyl bonds were sufficiently stable to allow the recovery of intact aminoacyl-tRNA from the column effluent.

Bernardi (1969a) investigated the elution profiles of ribosomal RNA from Ehrlich Ascites tumour cells, *E. coli* tRNA, viral RNA and polynucleotides using gradient elution. Single-stranded randomly-coiled polynucleotides were eluted at lower phosphate molarities than were double-stranded rigid polynucleotides, the latter always eluting at 0·2 to 0·22 M phosphate. A triple-stranded polynucleotide, consisting of two poly U and one poly A chain, eluted at a higher phosphate molarity (0·45 M).

DNA fractionation

Semenza (1957) first used HA columns to fractionate DNA and reported that variations in base composition could not be differentiated by this method. This was substantiated by Main, Wilkins and Cole (1959) who, using a modified method to prepare HA, fractionated a mixture of calf thymus DNA and polyadenylic acid using stepwise elution. The DNA was separated from the polyadenylic acid, and fractionated into four peaks, which eluted at 0·1, 0·12, 0·14 and 0·16 M phosphate. A small amount of polyadenylic acid did, however, contaminate the DNA (Figure 5.17).

Bernardi (1961) also fractionated calf thymus DNA using stepwise elution, and found that DNA (unlike RNA) is not degraded during chromatography as judged by light scattering, S-value and u.v. spectroscopy. Only two peaks were found which eluted at 0·2 and 0·25 M phosphate. Rechromatography of each individual peak, however, resulted in most of the material eluting at 0·2 M phosphate (Figure 5.18). The relative intensities of the two peaks depended on the length of the column, the first peak being larger than the second for short columns, while the reverse was true for longer columns. These artifacts could be overcome by using linear gradient elution, when only one peak was observed eluting at 0·2 to 0·22 M phosphate. No fractionation with regard to base composition was found, but high molecular weight DNA from T_5 and T_2 bacteriophages was eluted at a slightly increased phosphate molarity of 0·25 to 0·27 M (Bernardi, 1969b).

Bernardi (1969b) observed that DNA from bacteriophage Ø-X174 chromatographed in a different position to that of either yeast mitochondrial DNA or glycosylated DNA from T-even bacteriophages, and could only be fractionated by stepwise elution. Moreover, unlike calf thymus DNA, one peak only was obtained.

The chromatographic behaviour on HA of denatured DNA differs from that of native DNA. The main peak of denatured DNA elutes at 0·15 M phosphate, with small subsidiary peaks eluting at higher molarities (Bernardi, 1962, 1969c). This enables native and denatured DNA to be resolved completely (Chevallier and Bernardi, 1968) as shown in Figure 5.19.

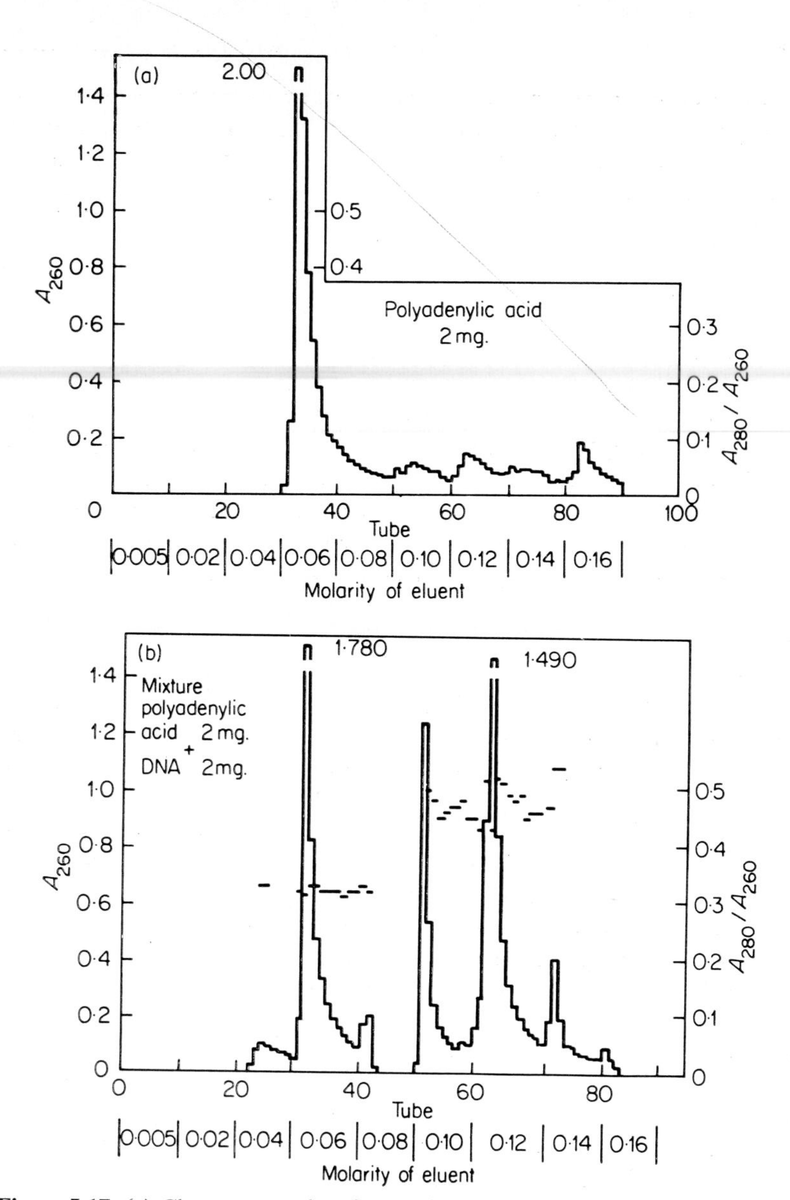

Figure 5.17. (a) Chromatography of polyadenylic acid (2.0 mg. dissolved in 4.0 ml. of 0·005 M phosphate buffer pH 6.7 placed on column at rate of 4·0 ml. per 45 min.) on hydroxyapatite column (10 × 75 mm.); temperature 6°. Elution by discontinuous gradient elution. Eluent, sodium phosphate buffers, pH 6.7, at molarities indicated. Eluent flow rate maintained at 4·8 ml./h. Optical absorbance of eluates (A_{260}) plotted according to scale at left. Values of bars (A_{280}/A_{260}) measured on scale at right. (b) Chromatography of mixture of polyadenylic acid (2·0 mg.) with calf thymus DNA (2·0 mg.) on calcium phosphate, CPM, column (10 × 75 mm.). Conditions of placement of mixture (solvent volume 4·0 ml.) on column and elution from column, same as in (a). Optical absorbancy data plotted as in (a). (After Main, Wilkins and Cole (1959) *J. Amer. Chem. Soc.* **81,** 6490. Copyright 1959 by the American Chemical Society. Reprinted by permission of the copyright holder)

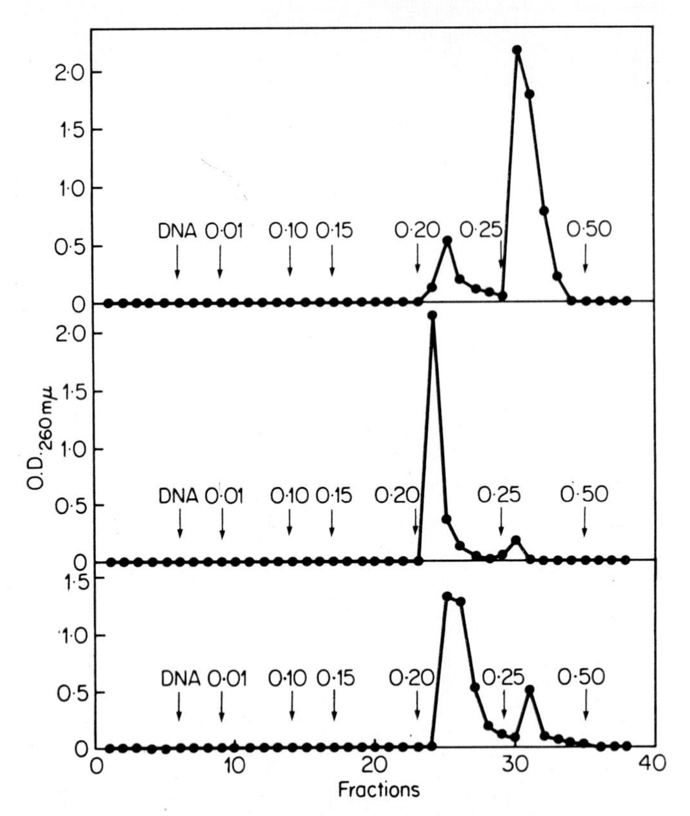

Figure 5.18. Top: chromatography of calf thymus DNA on hydroxyapatite. 1·28 mg. DNA A1; 1·3 × 5 cm. hydroxyapatite; 3 ml./fraction. Middle and bottom: rechromatography experiments on fraction a and b (pooled from two runs), respectively. 1·3 × 3 cm. hydroxyapatite; 3 ml./fraction. Middle: fraction a; bottom: fraction b. The stepwise increase in buffer molarity is indicated by the vertical arrows. (After Bernardi (1961))

When biologically active DNA is denatured, a small amount of residual transforming activity is invariably found associated with the denatured DNA. The origin of this activity has been the subject of considerable interest to many investigators. Chevallier and Bernardi (1965, 1968) and Bernardi (1969c) investigated this phenomenon using *H. influenzae* DNA. The DNA was denatured and fractionated on a HA column using gradient elution. The bulk of the denatured DNA was eluted at a phosphate molarity distinctly lower than that of native DNA. However, a small

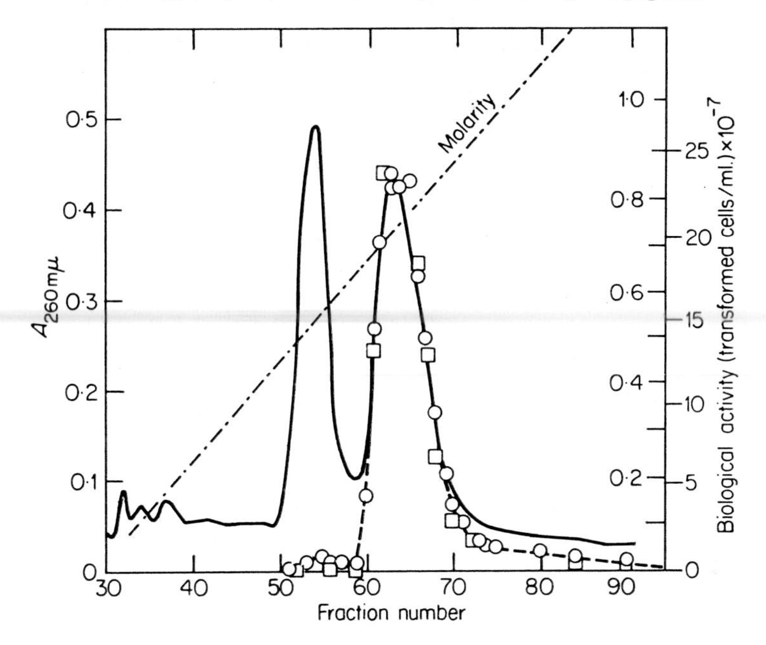

Figure 5.19. Chromatography of a mixture of native (streptomycin marker) and heat-denatured DNA isolated from *H. influenzae* (cathomycin marker). 31 ml. of a DNA solution (25 μg./ml.) were heat denatured, added to 5 ml. of a native DNA solution (75 μg./ml.) and loaded on a 2 cm. × 15 cm. hydroxyapatite column. Elution was carried out with a linear molarity gradient (100 ml. + 100 ml.) of potassium phosphate (0·001 to 0·5 M); 2·7 ml. fraction were collected. Circles indicate the cathomycin activity (right-hand, inner scale); squares indicate the streptomycin activity (right-hand outer scale). The elution molarity of the first peak was 0·14 M; that of the second peak, 0·21 M. Recovery of the optical density was 76 per cent; recovery of streptomycin activity, 76 per cent; recovery of cathomycin activity 62 per cent. Biological activity was tested at a DNA concentration of 0·05 μg./ml. (After Chevallier and Bernardi (1968))

amount of 'native-like' material eluted at the same molarity as native DNA, and carried the residual transforming activity for cathomycin marker (Figure 5.20). This 'native-like' DNA constitutes about ten per cent of the total denatured DNA and was found to be similar, although not identical, to native DNA with respect to certain physical characteristics. It was therefore concluded that the 'native-like' DNA molecules were distinct from native DNA molecules and postulated to have regions of single-strandedness in an otherwise double-helical structure. It is possible that the 'native-like' DNA molecules consist of denatured strands held together by covalent crosslinks, or alternatively possess the ability to renature very rapidly.

Thomas, Kelly and Rhoades (1968) were able to isolate superhelical DNA from bacteriophages T_7 and P_{22} after infection of *E. coli.* The total DNA in each case was denatured at pH 12·5 with alkali, neutralized with an acidic buffer and fractionated on HA. The superhelical DNA molecules are easily renatured because the two chains are covalently linked (see p.61). Denatured single-stranded DNA is eluted first, while the superhelical DNA molecules are initially retained by the column and eluted later.

The property of HA to discriminate between native double-stranded and denatured single-stranded DNA has been utilized by Miyazawa and Thomas (1965). These authors surrounded the HA columns with a water jacket so that the temperature could be accurately controlled (Figure 5.21). Sonicated short segments of wild type λ bacteriophage $cb2^+b5^+$ or *E. coli* DNA were loaded on to the column in 0·03 M phosphate buffer, and the column washed with starting buffer. The eluting buffer (0·08 M phosphate) was applied to the column which was equilibrated at a certain temperature for 5 min. and subsequently eluted. The process was then repeated at a different temperature. Since the T_m of a particular DNA molecule depends on its base composition, thermal chromatograms, which represent the amount of DNA melting at a given temperature, were constructed for different DNA's. A mixture of wild-type λ bacteriophage ($b2^+b5^+$) ^{14}C-labelled DNA and ^{32}P-labelled deletion mutant (b2b5) were compared in this way, and it was found that the ratio of $^{14}C/^{32}P$ increased above 1·0 on the low-temperature side of the profile. Taking into account suitable controls, Miyazawa and Thomas (1965) concluded that the sections of DNA which lacked the b2 and b5 genes were richer in AT composition (Figure 5.22). McCallum and Walker (1967) extended this technique to a study of segments of mouse and rat DNA, and found that certain sequences are distributed non-randomly among DNA fragments of differing base composition.

Bernardi and coworkers (1968) fractionated total nucleic acid extracts from both cytoplasmic 'petite' mutant and wild-type cultures of *S. cerevisiae* on HA column using a linear gradient of phosphate buffer. Five components were observed in each case. A small fraction eluted at the bed volume of the column and consisted of mono- and oligonucleotides. A second fraction, characterized as nuclear RNA, was eluted at 0·2 M phosphate. There were also two DNA components, designated fraction 'a' and fraction 'b' which eluted at 0·25 and 0·27 M phosphate, respectively. A small fifth component, fraction 'c', eluted at a high phosphate molarity (0·37 M) and had a very high GC content. The fractions

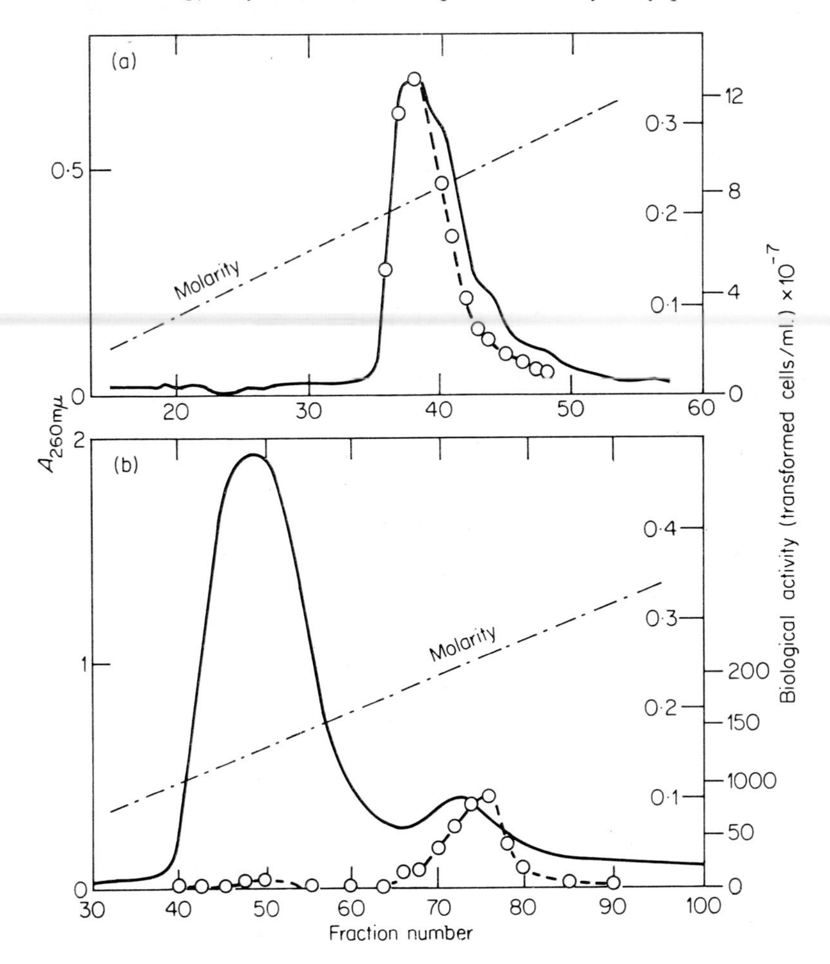

Figure 5.20. (a) Chromatography of native *H. influenzae* DNA: 1 ml. sample (835 μg. in saline phosphate) loaded on a 1 × 10 cm. HA column. Elution was carried out with linear gradient (100 ml. + 100 ml.) of potassium phosphate (0·001 to 0·5 M, inner scale); 3 ml. fractions were collected. Recovery of optical density was 85 per cent; recovery of biological activity (cathomycin marker; circles; right-hand outer scale) 82 per cent. (b) Chromatography of alkali-denatured *H. influenzae* DNA: 1850 ml. sample (1850 μg. in saline phosphate were loaded on a 1·2 × 20 cm. HA column. Elution was carried out with linear gradient (150 ml. + 150 ml.) of potassium phosphate (0·001 to 0·5 M, inner scale); 2·4 ml. fractions were collected. Recovery of biological activity (cathomycin marker; circle; right-hand outer scale) 98 per cent. The specific activity relative to undenatured DNA was 43 per cent for fraction 76, which exhibited the highest relative activity, and 29 per cent for fractions 66 to 85 (average value). (After Chevallier and Bernardi (1968))

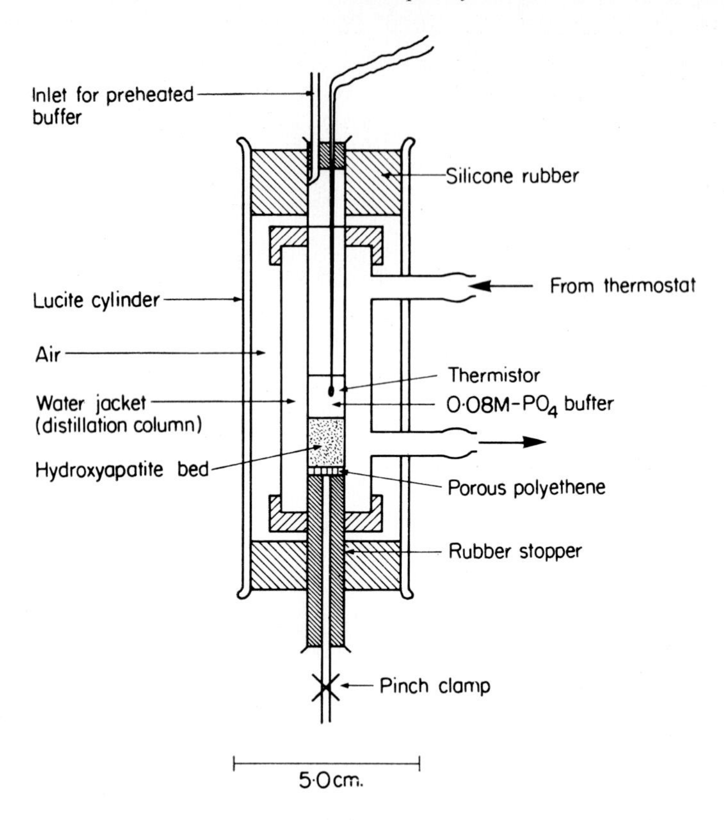

Figure 5.21. Jacketed chromatographic column. This device was used for thermal chromatography of sonic fragments of various DNA's. (After Miyazawa and Thomas, Jr., (1965))

were analysed using CsCl density-gradient centrifugation (see p. 56) and it was found that fractions 'a' and 'b' from wild-type cells had densities of 1·700 and 1·68g./cm.^{3} corresponding to nuclear and mitochondrial DNA, respectively. In the mutant cells, fractions 'a' and 'b' had densities of 1·700 and 1·672g./cm.^{3} corresponding to nuclear and satellite DNA, respectively. Electronmicroscopy of the satellite DNA (fraction 'b':10 per cent of total) showed both linear and circular molecules. The circles had a contour length of 0·5μ (corresponding to a molecular weight of 1×10^6 daltons) and the linear structures were about 1·0μ in length (molecular weight 2×10^6 daltons). Hydroxyapatite chromatography therefore provides a method of separating mitochondrial DNA from cellular DNA, suggesting that the two differ in their secondary structure.

Bourgaux and Bourgaux-Ramoisy (1967) isolated labelled DNA from mouse embryo cells infected with Polyoma virus, which on ultracentrifugation gave two components, I and II with S-values of 20 and 16, respectively. When a mixture of I and II was applied to a HA column and eluted with

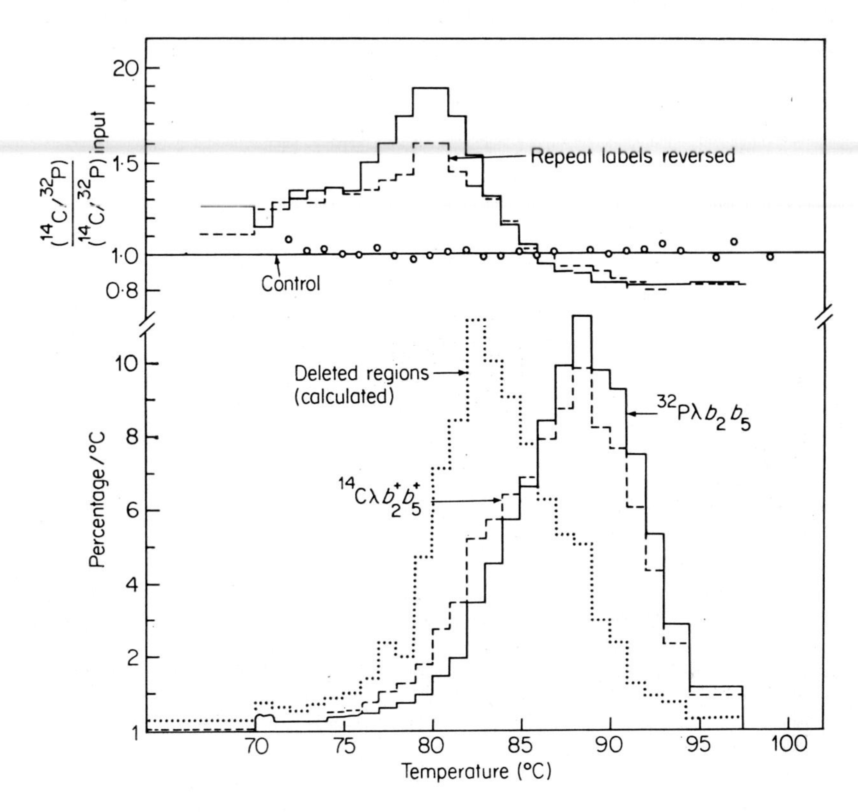

Figure 5.22. Thermal chromatograms of wild-type λ together with an alternately-labelled λ DNA having large deletion mutations. The distribution marked 'deleted regions' was calculated by taking the difference between the two thermal chromatograms as described in the text. It represents the best estimate of the thermal chromatogram of the regions of the λ DNA molecule which are missing in the b2b5 mutant. (After Miyazawa and Thomas, Jr., (1965))

a linear gradient of phosphate buffer, component I eluted prior to component II. Moreover, components I and II could be separated from the host cell DNA on HA columns (Figure 5.23). When the two components were subsequently heat-denatured and fractionated on a HA column, component II eluted at a low phosphate molarity, whereas the chromatographic properties of component I were largely unaffected. It was concluded that components I and II had different tertiary structures. Ultracentrifugation and electronmicroscopy showed component I to consist entirely of twisted circular (superhelical) DNA molecules, whereas component II consisted of a mixture of circular and linear molecules in variable proportions (Bourgaux-Ramoisy, van Tieghem and Bourgaux, 1967).

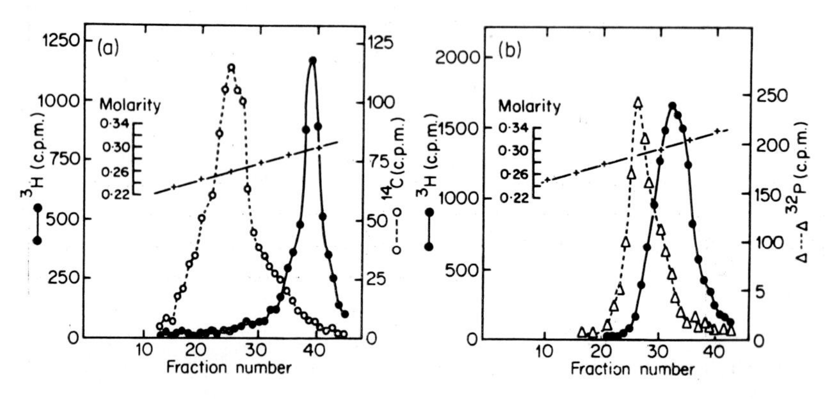

Figure 5.23. Chromatographic separation of viral and cellular DNA. A sample of ^{3}H-DNA from mouse embryo cells mixed with either component I labelled with ^{14}C (a) or component II labelled with ^{32}P (b) from polyoma virus DNA and fractionated by HA column. Linear gradient (0·23 to 0·32 M) sodium phosphate was used to elute the column and 0·5 ml. fractions were collected and assayed for radioactivity. (After Bourgaux and Bourgaux-Ramoisy (1967))

Hydroxyapatite can also be used to separate the protein and DNA components of soluble calf thymus nucleoprotein (Faulhaber and Bernardi, 1967). When the nucleoprotein is bound to and eluted from columns using linear phosphate gradients, only one peak at 0·2 M phosphate is observed. However, in the presence of 3·0 M KCl two peaks can be seen, one at 0·07 to 0·1 M phosphate containing protein, and one at 0·25 to 0·3 M phosphate comprising of DNA. Hence, in a high ionic strength medium, the two components of the nucleoprotein can be separated.

Rigid polynucleotides possessing a double-stranded structure are eluted from the HA columns at about 0·2 to 0·22 M potassium phosphate buffer. Slight discrepancies can occur which are probably related to their secondary structure, and this appears to explain the separation of nuclear and mitochondrial DNA's from yeast. Polynucleotides in the random-coil configuration are not bound to the column as firmly as those in the helical form, and are eluted at correspondingly lower phosphate molarities. It is possible that the ionic strength and temperature may influence the fractionation of single-stranded polynucleotides, by modifying their structure. A general explanation for this phenomenon could be that the phosphate groups, which were available for the interaction with the adsorbing sites (calcium ions) on HA in the rigid, ordered structure, greatly decrease in number on the 'outer surface' of the randomly-coiled denatured nucleic acids. There is no evidence to suggest that the bases or sugars of the polynucleotides play any role in this interaction.

Summary

Gel filtration on Sephadex or agarose gels, and polyacrylamide gel electrophoresis fractionate molecules of varying size by a molecular sieve effect. In the case of Sephadex and agarose, the small molecules are initially retained by the gel granules and eluted later, whereas the large molecules do not penetrate the granules and are eluted at the void volume. The opposite effect occurs with polyacrylamide gels, the smaller molecules being sieved more quickly than the larger. Polyacrylamide gel electrophoresis fractionates proteins according to their size and charge. The net charge of the various protein molecules is different because of the varying number and type of amino acid side chains present. All nucleic acids and polynucleotides, however, have a similar charge to mass ratio, and hence charge is not a determining factor in their fractionation.

The two molecular sieve methods have been used successfully to fractionate low and high molecular weight RNA (tRNA, mRNA and rRNA), viral RNA and small polynucleotides, and can be used to detect any degradation of a particular RNA species.

Gel filtration is a preparative method and the fractions can be obtained directly for further studies. On the other hand, polyacrylamide gel electrophoresis is largely an analytical method. Nucleic acids have been separated by preparative gel electrophoresis but the problems encountered make this a laborious task.

Nuclear DNA cannot be fractionated on either agarose or polyacrylamide gels. The gelatinous matrices restrict the flow of DNA solutions and this may lead to denaturation and degradation. DNA can, however, be separated from RNA on Sephadex gels since the former is eluted at the void volume. Moreover, it has been possible to separate the circular single-stranded and double-stranded forms of Ø-X174 DNA by polyacrylamide gel electrophoresis. In this case, single-stranded DNA may behave in a manner similar to RNA. It may be possible to separate circular R-factor DNA from chromosomal DNA using polyacrylamide gels since R-factor is circular and has a smaller molecular weight.

Polyacrylamide gel electrophoresis can be used to separate RNA species according to their base composition by altering the conditions of electrophoresis (for example, ionic strength and Mg^{2+} concentration) which are known to change the secondary structure of polynucleotides.

Fractionation on hydroxyapatite columns depends only upon the secondary structure of the nucleic acid species, and has been widely used to separate denatured (single-stranded) from native (double-stranded) DNA, and to fractionate circular (mitochondrial, viral and R-factor), linear and superhelical forms of native DNA. The method cannot be used to fractionate high molecular weight RNA as this is degraded by the process of chromatography. tRNA has, however, been resolved into various species but not as effectively as by ion-exchange chromatography (see Chapter 3). Hydroxyapatite may be used extensively in the future for DNA fractionation, since flow rates are good (50ml./h.) and recovery is excellent (95 to 100 per cent).

CHAPTER 6

Methylated-albumin and polylysine-kieselguhr chromatography

Introduction

The importance of basic nuclear proteins in the control of DNA metabolism and in chromosome structure has been well reviewed elsewhere (Hinilica, 1967). However, because of the structural heterogeneity of these proteins and the complexity of their interactions with DNA, very little is known about the nature and manner of the forces involved. Since the natural interactions are so complex many have studied model systems to try to throw some light on the subject (Tsuboi, Matsuo and Ts'o, 1966; Olins, Olins and von Hippel, 1967, 1968; Shapiro and coworkers, 1969). These studies led to the development of the techniques described in this chapter: Methylated-albumin- and polylysine-kieselguhr (MAK and PLK) chromatography which fractionate nucleic acids on matrices essentially consisting of basic proteins bound to the inert substance kieselguhr. It is therefore pertinent to discuss polycation–nucleic acid interactions of model systems to understand the mechanism of fractionation on MAK and PLK.

Polycation–nucleic acid interactions

The naturally occurring aliphatic polyamines spermine ($NH_2(CH_2)_3$-$NH(CH_2)_3NH_2$) and spermidine ($NH_2(CH_2)_3NH(CH_2)_4NH_2$) were investigated (Tabor, 1961, 1962; Mandel, 1962; Hirschman, Leng and Felsenfeld, 1967) for their interaction with DNA and, as more became known about the amino acid composition of the histones, the synthetic polyamino acids of arginine and lysine were also studied (Leng and Felsenfeld, 1966; Shapiro and coworkers, 1969). During the last three years attention has focused on the protamines (Olins, Olins and von

Hippel, 1968): these are the small proteins which occur in the place of histones in certain types of nuclei (Miescher, 1871). The substrate for these studies was usually native double-stranded DNA; however, synthetic polynucleotides have also been used, particularly when purity was important, and secondary forces due to the base composition of the nucleic acids were being investigated. Higuchi and Tsuboi (1966) also studied certain types of RNA.

As with basic proteins, polycations readily form insoluble complexes with DNA at low ionic strength. This system is more difficult to study, with a consequent diversity of experimental approaches. The manner in which the complex is formed, and its environment, are critical for the correct interpretation of the results.

Formation of the complex

Two main methods are used to form polycation-polynucleotide complexes: direct mixing and equilibrium dialysis.

Direct mixing usually involves addition of the polycation dissolved in dilute buffer, to a more concentrated solution of a polynucleotide, also in dilute buffer. Sober and coworkers (1966) adopted this procedure during their investigation of nuclease-resistant polylysine–RNA complexes and, in order to keep the complex soluble for enzymic study, the lysine : nucleotide ratio was maintained at 1 : 10. In experiments using polylysine, in which the ratio of lysine : nucleotide approaches 1 : 1, and the ionic strength of the buffer is still low, the complex formed precipitates out of solution. Leng and Felsenfeld (1966) added a polylysine solution to DNA in NaCl in a dropwise manner with vigorous stirring, and obtained such an aggregate. However, at higher NaCl concentrations, smaller amounts of DNA were precipitated and at M NaCl, the total amount of DNA precipitated from solution varied as a function of the DNA and polylysine concentration. Subsequently this method was standardized, and polylysine was added to solutions of DNA in 0·9 to 1·0 M NaCl, and incubated at 25° for 10–30min. Any precipitate was then removed by centrifugation at 25,000g. The solution of polylysine was always added in a dropwise manner with continuous vigorous mixing (Shapiro, Leng and Felsenfeld, 1969).

Direct mixing procedures at low ionic strengths were also used by Olins, Olins and von Hippel (1968) during their investigation into short-chain polypeptide–DNA interactions. In all cases, mixing resulted in immediate precipitation. However, after stirring for 16 h. at room temperature, certain solutions cleared, suggesting that equilibration had occurred. On the other hand, when larger polypeptide chains were used (lys. 200–350 residues), equilibrium complexes could not be formed by direct mixing. Another approach had to be used to obtain truly equilibrated complexes of long-chain polycations with polynucleotides at low ionic strength.

Huang, Bonner and Murray (1964) reconstituted nucleohistones by repeated dialysis of a solution containing the two components in high salt concentration, against decreasing concentrations of salt. This method was subsequently used to form polylysine–DNA complexes (Olins, Olins and von Hippel, 1967). Polylysine and DNA were mixed together in 4 M NaCl with vigorous shaking, at which ionic strength no precipitate was detected. Annealing of the complexes was performed by dialysis at 0°C against 0·001 M cacodylate buffer containing decreasing amounts of NaCl. Any precipitate that formed was removed by centrifugation at 1700g. In this way, complexes are initially formed under 'reversible' equilibrium conditions, and by progressive dialysis into a low salt concentration, the system is 'frozen' into 'irreversibility'. Using this method, it was hoped that the final state achieved more closely represented the interaction specificity (and equilibrium) which applies at low salt concentration, but which cannot be attained by direct mixing.

Stabilization of the DNA double helix

The helix to coil transition of native DNA is effected by increasing the temperature. At a certain point, termed the T_m, which is characteristic of the molecule, the two DNA chains separate, and the resulting increase in hypochromicity can be monitored at 260mμ (see Chapter 2).

A number of simple cationic compounds stabilize native DNA against thermal denaturation, presumably by preferentially binding to the native DNA helix. These compounds include most of the mono- and divalent cations (Dove and Davidson, 1962; Schildkraut and Lifson, 1965), spermines, spermidines and diamines. In general, the spermines and spermidines are effective at much lower concentrations than divalent cations (Mahler, Mehrotra and Sharp, 1961), which are in turn more effective than monovalent cations.

Tabor (1961) found that 10^{-4} M spermine increased the residual transforming activity of DNA from *B. subtilis* after heat denaturation. Furthermore, the increase in T_m was about 8°C (Tabor, 1962). These results agreed with those of Mahler, Mehrotra and Sharp (1961), who investigated the effect of diamines on the thermal transition of DNA. Working with diamines of increasing chain-length between the amino groups, diaminopentane ($H_2N(CH_2)_5NH_2$) raised the T_m by approximately 5·3°, whereas diaminoethane and diaminooctane only increased the T_m by

1·3° and 0·7°, respectively. Thus, the distance between the amino groups was an important factor in the stabilizing effect.

The interaction of all these small cationic substances with native DNA produces a monophasic thermal transition, the T_m values of which depend on the concentration (up to saturation) of the cationic substance in solution. This effect can also be shown with oligotetralysine which behaves like salts, spermine, spermidine and diamines with respect to its dynamic stabilization of DNA (Olins, Olins and von Hippel, 1968). Thus, in this type of system, it would appear that all the DNA in solution is at least partially stabilized by the available cations, and that the binding is not irreversible, but that a state of dynamic equilibrium exists between the DNA and the available cations.

This dynamic behaviour is in marked contrast to the type of stabilization achieved by the cationic polypeptides at low ionic strength, where an essentially irreversible complex is formed, and the addition of more polypeptide (under annealing conditions) changes the amount of complex present, but not its melting temperature. Such studies have been carried out by Tsuboi, Matsuo and Ts'o (1966), who investigated the interactions of polylysine with poly IC and calf thymus DNA. Both types of complex gave a biphasic thermal transition, the first stage (lower temperature) corresponding to the T_m of the free polynucleotide, and the second (higher temperature) to the T_m of the complex. These transitions were independent of the polylysine concentration. However, as the amount of polylysine increased, the percentage of the total hypochromicity found at the first stage decreased, while a corresponding increase occurred at the second stage, and the complexed and free DNA moieties could be separated using isoelectric focusing. Olins, Olins and von Hippel (1967), using a polybasic amino acid and native calf thymus DNA, observed essentially similar biphasic melting profiles, and also monitored the turbidity changes at 350 mμ of the entire complex. With biphasic melting transitions, therefore, the first melting represents the helix to coil transition of free DNA molecules, whereas the second is characteristic of the particular DNA–polypeptide complex involved. An elevation in T_m reflects the degree to which that particular polypeptide has stabilized the DNA against thermal denaturation. Polypeptides therefore prevent the unwinding of the two DNA strands, thus raising the temperature required for denaturation. The naturally occurring protamines of fish sperm, which contain between 20 and 24 arginyl residues in single peptide chains of about 30 amino acid residues (Ando and Suzuki, 1966), strongly resemble the oligolysines 14–18 in their interactions with DNA (Olins, Olins and von Hippel, 1968). The thermal denaturation profiles of DNA–protamine complexes show that the first melting transition shifts towards the higher T_m values with increasing amounts of protamine. The relative effective polycation length in natural protamines is therefore shorter than that found in synthetic polypeptides, and closer to that of the oligopeptides so far investigated.

Forces of interaction

Matsuo, Mitsui, Ititaka and Tsuboi (1968) obtained an infrared spectrum of a stoichiometric complex between calf thymus DNA and polylysine. At 92 per cent relative humidity, the PO_2^- asymmetric stretching vibration was found at 1224cm.$^{-1}$, as with DNA alone. On drying to zero per cent relative humidity, the DNA complex gave a band at 1233cm.$^{-1}$, but pure DNA, gave a band at 1241cm.$^{-1}$. A shift of 1224 to 1241cm.$^{-1}$ occurred when water was removed, and the PO_2^- group became available for hydrogen bonding. If, however, a $NH_3^+ \ldots PO_2^-$ unit existed in the complex, this would remain even after the removal of water, causing only a small shift to 1233cm.$^{-1}$. Hence it was concluded that the NH_3^+ group at the end of the side chains of polylysine (i.e. the protonated ε-NH_2 group) was forming a hydrogen bond with the PO_2^- group. These results confirmed previous interpretations: there is a stoichiometric ratio of lysine : nucleotide of 1 : 1 in the complex formed with double-stranded DNA (Higuchi and Tsuboi, 1966), and the positively charged ε-amino groups on the polylysine side chains are interacting with the negatively charged phosphate groups on the polynucleotide molecules.

The structural conformation of a polycation–polynucleotide complex allowed by this stoichiometry was investigated by Liquori and coworkers (1967) using X-ray analysis. The shape and dimensions of the polyamine, spermine, are such that they allow stereospecific interaction with DNA, the polyamine lying in the narrow groove of the macromolecule. This interaction between oppositely charged polyelectrolytes, coupled with the preferred stereospecificity of the polyamine, forms the major force holding the two types of molecules together. However, earlier workers had realized that these were not the only forces involved (Spitnik, Lipshitz and Chargaff, 1955). Precipitates, formed by the reaction of DNA with histones or polylysine, could be redissolved in NaCl solutions of concentrations 1–2 M, and the base composition of each succeeding precipitated fraction was richer in the two bases adenine and thymine.

Lucy and Butler (1955) challenged these findings by investigating the proteins present in calf thymus nucleoprotein fractions. They postulated that the nucleoprotein contained some molecules in which nucleic acid with a high GC content was linked to a histone containing a high proportion of lysine, while other molecules contained DNA rich in AT and linked to arginine-rich histone. Secondary forces, which varied according to the base composition of the DNA and the nature of the polyamino acid side chain, were therefore important in directing complex formation, but their exact nature was not revealed.

To elucidate the interaction between DNA's of varying base composition and polylysine and polyarginine, Leng and Felsenfeld (1966) studied complexes formed under reversible conditions. In M NaCl, polylysine showed selectivity for AT-rich DNA, while polyarginine exhibited a slight preference for DNA with a GC content of 72 per cent. However, the ionic conditions were also important, since selectivity was poor in 0·1 M rather than 1 M NaCl. There was little to suggest the mechanism involved in such specific interactions.

Wagner and Arav (1968) studied the interaction and binding of 5′-ribo- and 5′-deoxyribonucleoside monophosphates by polylysine and polyarginine using equilibrium dialysis. The binding behaviour depended on the nature of the base, and in every case guanylic acid bound most strongly. These results agree closely with those of Sober and coworkers (1966), who reacted polylysines (on an equivalent basis) with RNA to form insoluble complexes at low salt concentration. The use of non-specific ribonucleases on the soluble complex, resulted in a precipitate with a lysine : nucleotide ratio of 1 : 1, and a 'protected' nucleotide chain of essentially the same length as the polylysine of the initial complex. The protection specificity was for GC-rich fragments, and thermal denaturation markedly reduced this specificity. Their data were consistent with a model in which the polylysine molecule was protecting a short length of double-stranded RNA in the form of a 'hairpin' coil. Subsequently, Latt and Sober (1967a) carried out binding studies using individual oligomers of the (L-lysine)$_n$-ε-*N*-DNP-L-lysine series ($n = 3$ to 8) and the synthetic polynucleotides poly (I + C) and poly (A + U). In soluble complexes, binding was stronger to poly (I + C) than to poly (A + U), and both the total binding energy, and the difference between the binding energies to poly (I + C) and poly (A + U), increased linearly with oligolysine chain-length.

The competitive effect of monovalent and divalent cations on the protein–nucleic acid interaction was also studied (Latt and Sober, 1967b). Na^+ had (AU) or (AT) specificity, and tended to displace the basic protein to (IC)- or (GC)-rich regions, provided cooperative effects or aggregation (occurring after binding had reached a certain level) did not cause the weakly bound regions to aggregate preferentially. It was thought initially that the interaction of spermine with DNA preferentially protected AT-rich regions in a similar manner to polylysine (Mandel, 1962), but this conclusion still lacks proof (Hirshman, Leng and Felsenfeld, 1967).

Secondary forces influence the binding of the polycations to the polynucleotides, but their exact nature is uncertain. Experimental conditions are important, and the interpretation of any results should take into account the following points: ionic strength of the medium, reversibility of association and disassociation, the nature of the polycation base, the length of the polycation chain, the stoichiometric ratio of cation : nucleotide and the solubility of the complex.

Several of these are interrelated: for example, the reversibility of association is related to the ionic strength of the medium, and the stoichiometric ratio of the cation : nucleotide affects the solubility of the complex. Moreover, variation in any one parameter will affect the binding forces of the complex, and must be carefully considered when interpreting any results. It is also possible that the forces involved in aggregate formation are predominantly different to those which cause the initial binding or nuclease protection: the masking of one effect by another should therefore be considered. However, despite these uncertainties, the ease with which polylysine and the nucleic acids associate and disassociate with changing ionic strength is extremely useful, and provides the basis of polylysine-kieselguhr chromatography.

(A.) Methylated albumin-coated kieselguhr (MAK)

Preparation of MAK column

The method of preparing MAK columns usually follows the original methods of Fraenkel-Conrat and Olcott (1945) and Mandell and Hershey (1960) and involves the preparation of methylated-albumin followed by the protein-coated kieselguhr. Albumin (Fraction V bovine serum albumin, 5g.) is dissolved in absolute methanol (500ml.) and 12N HCl (4·2ml.) added. The protein first dissolves, but precipitates on standing for three days or longer in the dark with occasional shaking. The esterified albumin precipitate is collected by centrifugation, washed twice with methanol, twice with anhydrous ether and dried in vacuo over KOH.

Kieselguhr (20g.) suspended in 0·1 M buffered saline, pH 6·8 (100ml.) is boiled (to expel air) and cooled and a solution of 1 per cent esterified albumin (5ml.) added with stirring. The mixture is then washed with 0·1 M buffered saline (300ml.). The column is prepared by packing a known amount of MAK under an air pressure of 2–3lb/sq.in. until the excess buffered saline reaches the level of the packed material. For stepwise elution, Sueoka and Cheng (1962) used columns in which the height of the packed material was between 1 and 2cm. However, Mandell and Hershey (1960) used columns of three layers in which heavily-coated, lightly-coated and uncoated kieselguhr were packed into a column in order. This prevented channelling by allowing weak interactions (between the nucleic acid and uncoated kieselguhr) to take place first. The column is then washed with buffered saline (10 volumes to 1 volume of MAK). In general, the salt concentration is 0·4 M NaCl, but for smaller DNA or RNA, 0·1 M buffered saline is used. The nucleic acid sample is adjusted to a concentration of about 20μg./ml. and loaded under an air pressure which gives a flow rate of about 3ml./min. Elution is either by a gradient or stepwise increase of salt molarity, starting with the same concentration used for equilibrating the column and ending with M NaCl.

Biochemical applications of MAK chromatography

MAK chromatography was first used by Lerman (1955) to fractionate streptomycin resistant pneumococcal transforming DNA. The methylated bovine serum albumin was prepared by the method of Fraenkel-Conrat and Olcott (1945), and mixed with 20 parts of celite. Columns packed with this preparation were loaded with ^{32}P DNA, and washed with 0·1 M NaCl before eluting with an increasing NaCl gradient. Fractions were collected and assayed for transforming activity, and the highest specific transforming activity observed was approximately twice that of the original DNA. Lerman (1955) concluded that the formation of a complex between DNA and the basic protein displaced the cations and anions which were bound to DNA and protein, respectively, and that elution was determined not only by the ionic strength of the medium, but by the competition of its ions for the interacting charged sites of both the DNA and the protein. The technique was investigated further by Mandell and Hershey (1960) who found that gradient elution would separate T_2 from T_4 phage DNA (Figure 6.1). Shearing resulted in some of the DNA molecules eluting earlier, while heat denaturation caused their irreversible binding to the MAK, and even 3 M NaCl could not elute them. Hershey and Burgi (1960) used MAK to measure the breakage of T_2 phage DNA caused by stirring, and concluded that the change in chromatographic behaviour was due to single breaks near the centre of the molecules (Figures 6.2 and 6.3).

The fractionation of RNA on MAK columns was studied by Mandell and Hershey (1960) and also by Ageno and coworkers (1966), who separated tRNA from DNA and from ribosomal RNA (Figure 6.4). Their results were in agreement with those obtained by Cocito, Prinze and De Somer (1961) during their fractionation of infectious poliovirus RNA on MAK columns. By slightly altering the gradient, ribosomal RNA could be further separated into two peaks, corresponding to the

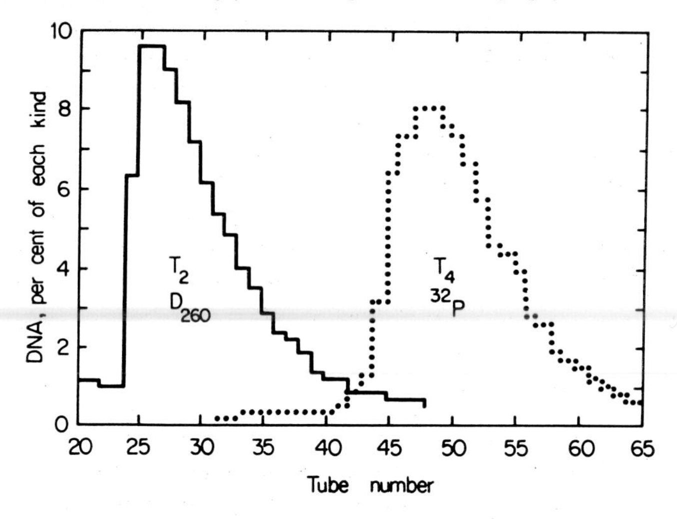

Figure 6.1. Elution pattern obtained with a mixture of 0·3 mg. of unlabelled T_2 DNA and tracer amounts of ^{32}P-labelled T_4 DNA. NaCl concentration gradient 0·0012 M per tube (3·6 ml.); concentration at T_2 peak, 0·72 M. Tubes are numbered from the gradient front. (After Mandell and Hershey (1960))

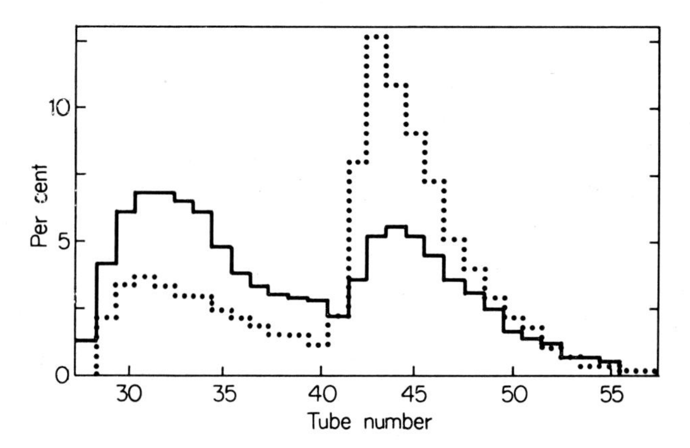

Figure 6.2. Measurement of breakage. Dotted line, ^{32}P-labelled DNA stirred for 2 min. at 260 rev./min. and 0·4 μg./ml., 30 per cent broken. Solid line, unlabelled DNA stirred for 15 min. at 8000 rev./min. and 0·4 mg./ml. 60 per cent broken. (After Hershey and Burgi (1960))

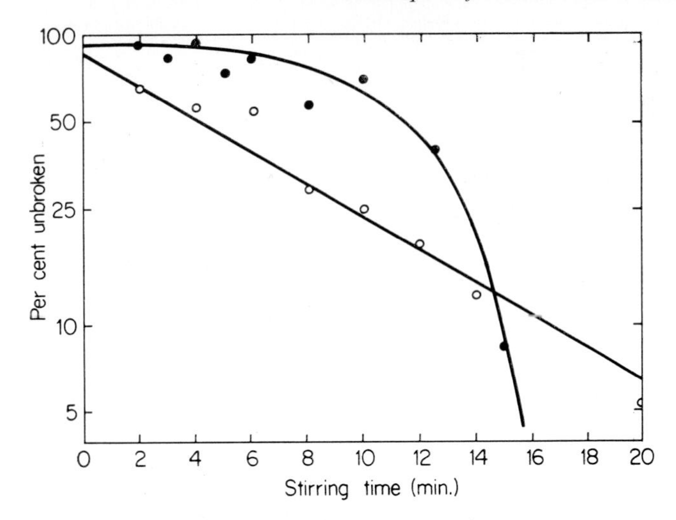

Figure 6.3. Kinetics of breakage. Open circles: DNA stirred at 260 rev./min. and 0·4 μg./ml. Each point is the average of 2 independent experiments. Filled circles: DNA stirred at 500 rev./min. and 4 μg./ml. Measurements as in Figure 6.2. (After Hershey and Burgi (1960))

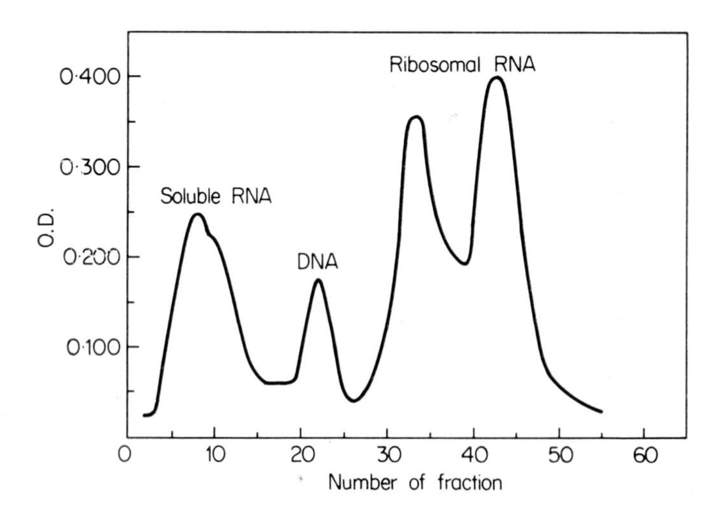

Figure 6.4. Elution pattern from a methylated-albumin kieselguhr column for a typical RNA preparation. Saline gradient was between 0·2 and 0·8 M NaCl, O.D. is optical density measured at 2600 Å. (After Ageno and coworkers (1966))

16S and 23S RNA of the ribosomal sub-units (Philipson, 1961). The method was simplified by Sueoka and Cheng (1962) who adopted a stepwise elution procedure which revealed additional properties of the MAK column. tRNA could be readily separated from the other nucleic acids, and 16S and 23S ribosomal RNA were in different fractions (Figure 6.5).

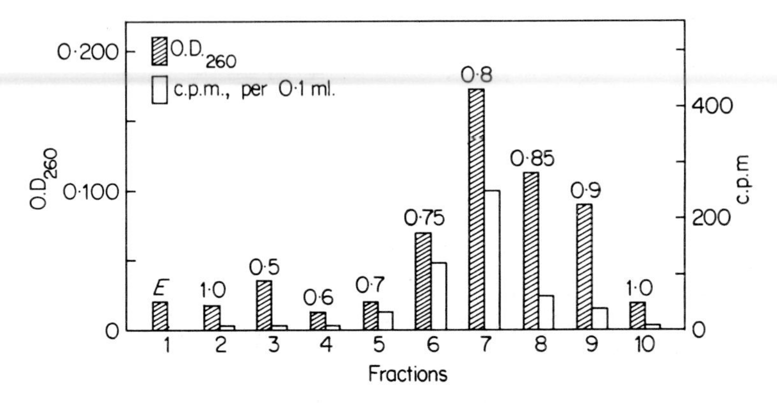

Figure 6.5. Fractionation of 16S and 23S ribosomal RNA. A sample of 16S ribosomal RNA which had been isolated by sucrose gradient method from [^{3}H] uridine labelled *E. coli* cells was mixed with total RNA from unlabelled *E. coli* cells. (After Sueoka and Cheng (1962))

The effect of base composition and hydrogen-bond content of DNA was also studied, and GC-rich DNA was eluted at a lower salt concentration than AT-rich DNA. Furthermore, thermally denatured DNA was eluted at a higher salt concentration than was native DNA. Cheng and Sueoka (1962) made further complementary studies by investigating the MAK fractions of DNA for heterogeneity in base composition. Using the caesium chloride density-gradient technique (Meselson, Stahl and Vinograd, 1957), it was found that fractions eluting at higher NaCl concentrations had lower buoyant densities. Since the buoyant densities of DNA in caesium chloride are linearly related to the mole per cent GC (Sueoka, Marmur and Doty, 1959) it was concluded that the fractions eluting first had a higher GC content (Figure 6.6).

These early experiments indicated the basis of the fractionation technique: the nucleic acids are attracted electrostatically to the basic protein on the kieselguhr and, as the ionic strength of the eluting buffer is increased, competition takes place between the ions and the charged groups of the nucleic acid and protein, causing the nucleic acid to be

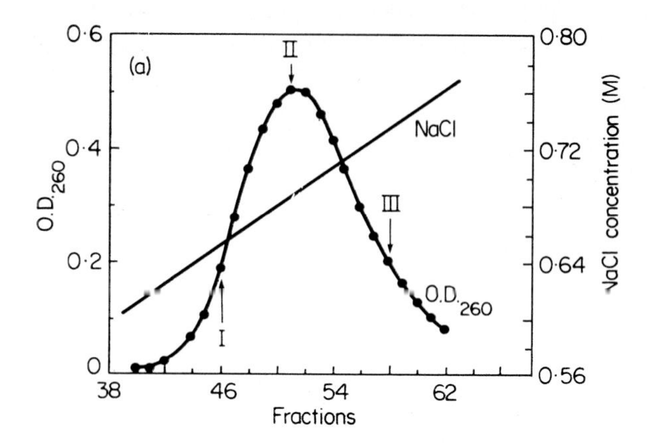

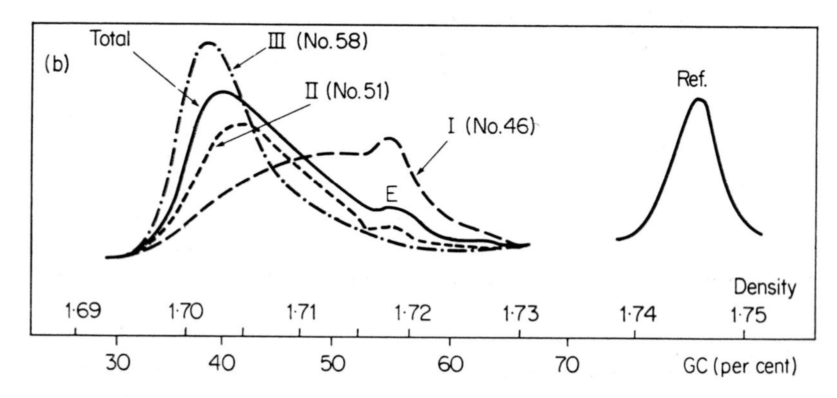

Figure 6.6. Elution profiles of DNA from a methylated albumin column. (a) Calf-thymus DNA; 1 mg. of calf-thymus DNA was dissolved in 50 ml. of saline-citrate solution and charged on a methylated albumin column. Fractionation was made by a linear gradient of NaCl between 0·5 M and 0·9 M in 0·05 M sodium phosphate buffer (pH 6.7). The volume of the starting and the final buffers was 100 ml., and 3-ml. fractions were collected. (b) Superimposed tracings of density-gradient centrifugation pictures of fractionated DNA from calf thymus. Samples of DNA (1 to 2 μg.) from front (I), middle (II) and rear (III) fractions (Figure 6.6a) were centrifuged separately in 7.7 molal CsCl solution at 44,770 rev./min. for 20 h. at 25°C. As a density reference, 1 μg. of ^{15}N-DNA from *Ps. aeruginosa* was added. Tracings of u.v. absorption pictures were superimposed by matching the DNA reference band. Distribution of unfractionated DNA (total DNA) was also added to the figure. E, extra component; ref., reference DNA (^{15}N-DNA from *Ps. aeruginosa*). (After Cheng and Sueoka (1962) *Science*, **141,** 1194. Copyright 1962 by the American Association for the Advancement of Science)

eluted. Transfer RNA is eluted first, (after nucleotides and small oligonucleotides) at approximately 0·4 M NaCl, followed by DNA. The secondary structure, molecular weight and the base composition of the DNA molecule also influence the position of RNA elution. In general, a decrease in molecular weight, or increase in GC content results in a molecule being retained by the column until the concentration of 1 M is reached. Ribosomal RNA elutes at 0·8 and 0·9 M NaCl, and the 16S molecule usually precedes the 23S molecule (Figure 6.7).

Fractionation of transforming DNA

MAK chromatography has now been used to solve many problems associated with nucleic acid research, and one of the earliest uses was in the fractionation of transforming DNA. Roger and Hotchkiss (1961) found that different genetic markers of pneumococcal transforming DNA are inactivated in solution at characteristic temperatures, analogous to melting points. These temperatures are sufficiently distinct as to permit the inactivation of the markers melting at lower temperatures, while leaving the remaining markers intact. Roger (1964), using stepwise elution, was able to separate the biologically active and inactive fractions of pneumococcal DNA and show that in the fractions eluting from MAK where native DNA was expected, the specific transforming activities for the preserved markers were substantially greater than those for untreated native DNA.

Saito and Masamune (1964) fractionated *B. subtilis* strain Marburg DNA on MAK using a linear gradient of 0·64 to 0·80 M NaCl, and recovered about 40 per cent of the applied DNA. A further 20 per cent of the DNA was eluted at 1 M NaCl and exhibited transforming activities similar to unfractionated DNA. The transforming activities of the DNA fractions were tested and, although the resolution was poor, the results indicated that enrichment of certain genes had taken place in certain fractions. A higher degree of resolution was achieved by Ayad (1964) and by Ayad, Barker and Jacob (1966), who fractionated *B. subtilis* DNA using a MAK column eluted with a linear gradient of NaCl-phosphate buffer (0·64–0·8 M NaCl), and obtained three peaks (Figure 6.8a). Fractions from the first and third peaks were pooled separately, dialysed against 0·64 M NaCl and refractionated on MAK (Figures 6.8b and c). Each fraction was adjusted to a final concentration of 1 μg./ml. and tested for transforming activity using competent cells of *B. subtilis* strain 3115 (histidine$^-$, tryptophan$^-$ and arginine$^-$), by the method described previously (see Chapter 1). It was found that a partial separation of the arginine, histidine and tryptophan genes had been achieved (Figure 6.8c).

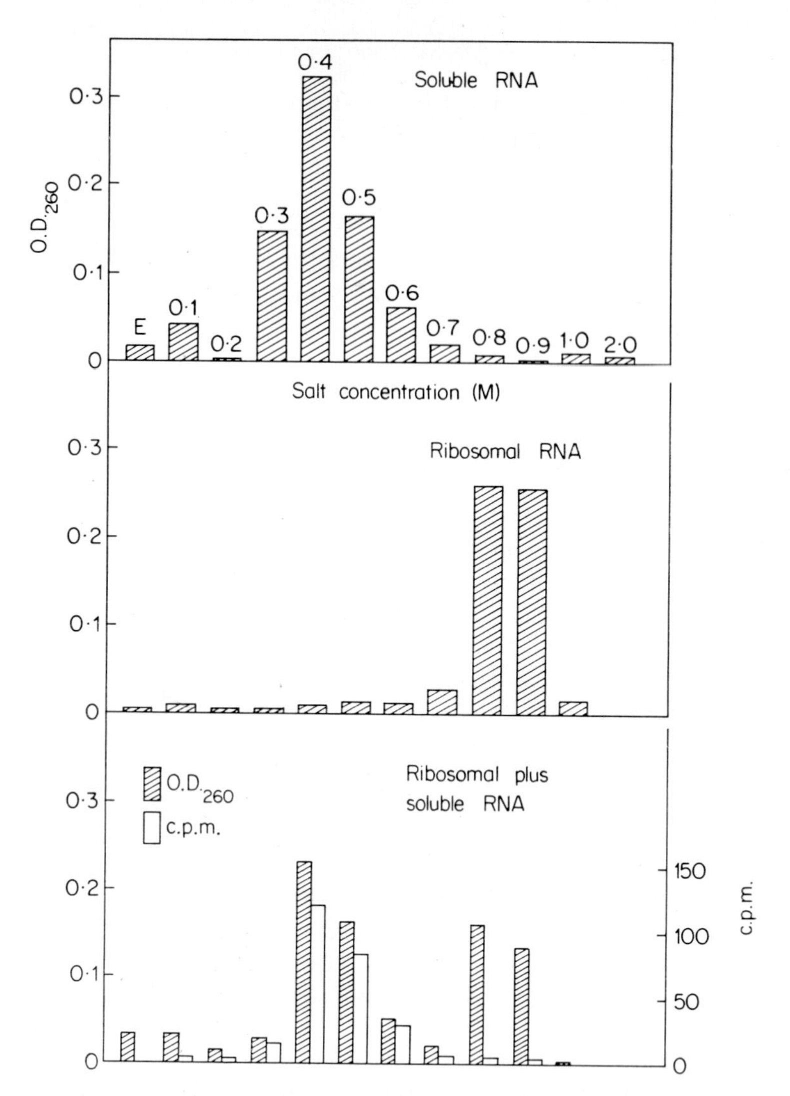

Figure 6.7. Elution pattern of soluble and ribosomal RNA, and their mixture from *E. coli*. In the case of the mixture soluble RNA was labelled with tritiated uridine. The method for counting radioactivity described by Hall and Spiegelman (1961) was followed. A portion (0·5 ml.) of each fraction was mixed with 200 μg. herring sperm DNA (0.1 ml.) and the nucleic acid was precipitated with 0·6 ml. of 20 per cent trichloroacetic acid (TCA). After 10 min. at 0°C, 1·2 ml. of 10 per cent TCA was added. The precipitate was collected and washed on a millipore filter (coarse, 27 mm. diameter). The filter was dried in a 50°C oven for 2 h. and placed in a glass vial filled with 15 ml. redistilled toluene containing 1·5 mg. of 1,4-bis-2-(5-phenyloxazolyl)benzene (POPOP) and 60 mg. of 2,5-diphenyloxazole (PPO). Counting was made in a Packard Tri Carb liquid scintillation counter. (After Sueoka and Cheng (1962))

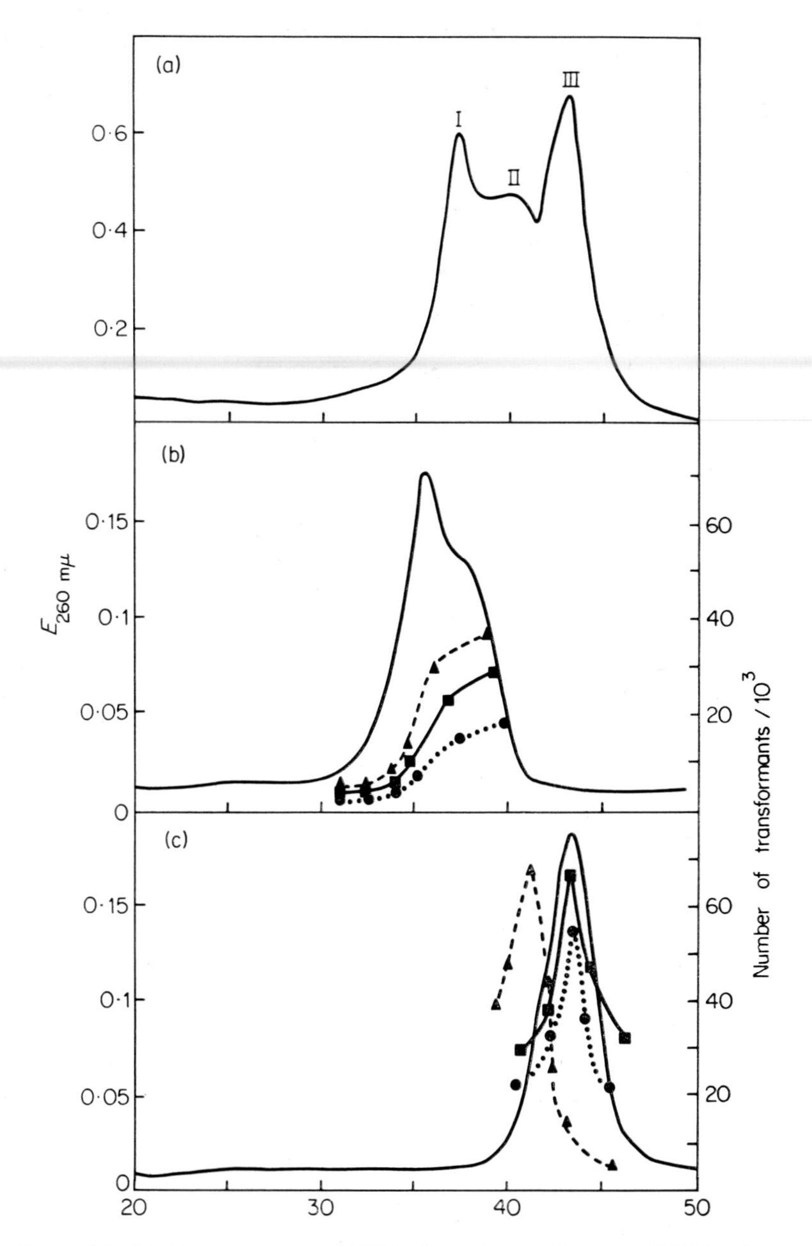

Figure 6.8. (a) Fractionation of DNA from *B. subtilis* using MAK column. 1 mg. of DNA was loaded on to the column and eluted with a linear gradient between 0·64 and 0·8 M-NaCl buffered with 0·05 M sodium phosphate pH 6.7. E_{260} of the eluate was measured continuously by an LKB Uvicord u.v. spectrophotometer. (b) Rechromatography of peak I. (c) Rechromatography of peak III. Fractions obtained from (b) and (c) were assayed for transforming activity using competent cells of *B. subtilis* strain 3115 (histidine gene, ●…●; tryptophan gene, ■—■; arginine gene, ▲—▲). (After Ayad, (1964))

Native DNA from *B. subtilis* has also been fractionated using stepwise elution from MAK (Ayad, Barker and Weigold, 1968). The transforming activity was confined to two out of the four fractions obtained, and a partial separation of DNA active in transformation for the arginine, histidine and tryptophan markers was achieved. Denaturation of the DNA at 100°C, followed by chromatography on MAK, yielded five fractions, two of which (eluted at 0·8 and 0·85 M NaCl, Figure 6.9) exhibited residual transforming activity (Table 6.1). Redenaturation at 100°C, followed by CsCl density-gradient centrifugation, resulted in the interconversion of four out of the five fractions. However, redenaturation of fractions labelled with ^{15}N and ^{2}H suggested the presence of a specific component that did not readily take part in the interconversions.

Stepwise elution of denatured pneumococcal DNA from MAK was also carried out by Roger, Beckmann and Hotchkiss (1966). Four fractions were obtained at different eluting salt molarities (Figure 6.10), two of which (fractions III and IV) regained high biological activity on annealing. Unfortunately the resolution was poor, but it was considered that the two fractions represented the complementary strands of the pneumococcal DNA. Roger (1968) attempted to improve the resolution by using gradient elution and maintaining the column at low temperature (Figure 6.11). She was able to show two peaks which were incompletely resolved, but which rechromatographed as distinct entities. Neither of the two peaks, however, possessed significant biological activity when renatured separately, but the central overlapping region of the two peaks showed a 30 per cent increase in biological activity when renatured. If fractions from both peaks were renatured together, 52 to 55 per cent of the transforming activity of native DNA could be recovered.

The use of intermittent gradients

The two strands of DNA are incompletely resolved on MAK. Rudner, Karkas, and Chargaff (1968a), however, overcame this difficulty by using an intermittent gradient: on the appearance of the first peak, the gradient was interrupted and restarted only when the top of the peak had been eluted. In this way, the eluting salt molarity remains constant (as in stepwise solution) until the first component has been liberated, and is then allowed to continue increasing linearly until the second component begins to appear (Figure 6.12). Rudner and coworkers were therefore able to separate, without any overlap, the two strands of *B. subtilis* DNA, which they designated H (heavy) and L (light) because of their buoyant densities in caesium chloride. The transforming activities, the temperature-absorbance profiles and the buoyant densities were recorded for both

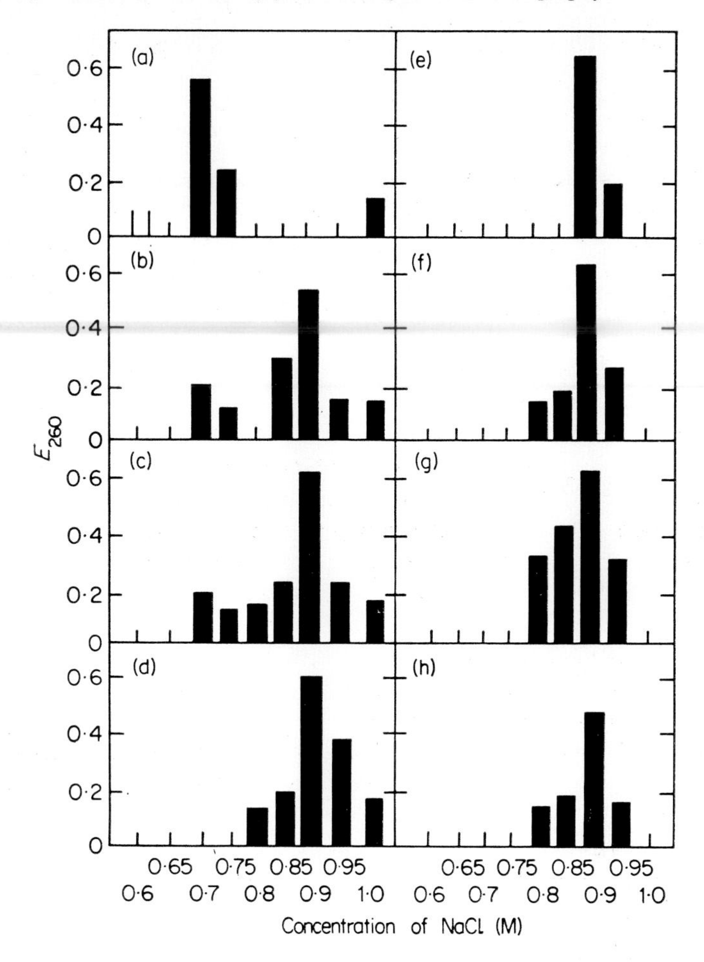

Figure 6.9. DNA from *B. subtilis* was fractionated by elution from MAK column with increasing concentrations of NaCl. E_{260} of the eluate was measured continuously by an LKB Uvicord u.v. spectrophotometer. (a) Native DNA; (b) DNA denatured at 90°C; (c) DNA denatured at 93·5°C; (d) DNA denatured at 100°C; (e) DNA eluted with 0·7 M NaCl, Figure 6.9a denatured at 100°C in the presence of formaldehyde; (f) DNA eluted with 0·7 M NaCl, Figure 6.9a, denatured at 100°C; (g) DNA eluted with 0·8 M and 0·85 M NaCl, Figure 6.9d, denatured at 100°C; (h) DNA eluted with 0.9 M and 0·95 M NaCl, Figure 6.9d, denatured at 100°C. (After Ayad, Barker and Weigold, (1968))

Table 6.1 Transforming activities of samples of DNA prepared from the Marburg strain of *B. subtilis*[a]

DNA sample	*Elution diagram*	*Conc. of NaCl in eluted fractions* (M)	*Plateau transforming activities, expressed as percentages of the value with an unfractionated native DNA control*		
			Arginine marker	*Histidine marker*	*Tryptophan marker*
Native	Fig. 6.9(a)	0·60	0	0	0
		0·70	161	82	71
		0·75	63	157	141
		1·0	0	0	0
Denatured at 90°	Fig. 6.9(b)	Unfractionated	89	71	66
		0·70	121	15	7
		0·75	23	85	80
		0·85	0	7	7
		0·90	0	0	0
		0·95	0	0	0
		1·0	0	0	0
Denatured at 93·5°	Fig. 6.9(c)	Unfractionated	57	8	11
		0·70	60	0·5	0·3
		0·75	1	8	7
		0·80	9	0	0
		0·85	0	4	4
		0·90	0	0	0
		0·95	0	0	0
		1·0	0	0	0
Denatured at 100°	Fig. 6.9(d)	Unfractionated	0·6	0·7	0·8
		0·80	0·7	0	0
		0·85	0	0·7	0·6
		0·90	0	0	0
		0·95	0	0	0
		1·0	0	0	0
0·8M- and 0·85M- NaCl fractions (Fig. 6.9d) heated at 100°	Fig. 6.9(g)	0·80	0·5	0	0
		0·85	0	0·7	0·7
		0·90	0	0	0
		0·95	0	0	0
0·9M- and 0·95M-NaCl fractions (Fig. 6.9d) heated at 100°	Fig. 6.9(h)	0·80	0·4	0	0
		0·85	0	0·8	0·8
		0·90	0	0	0
		0·95	0	0	0

[a]After Ayad, Barker and Weigold (1968).

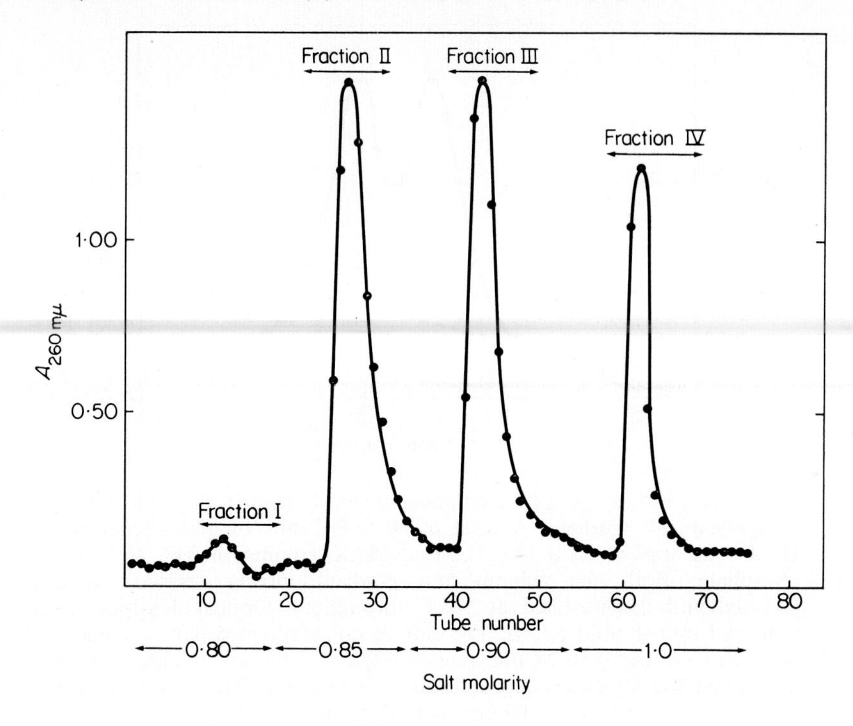

Figure 6.10. Fractionation of denatured pneumococcal DNA. 230 ml. of a solution containing 40 μg./ml. DNA in 0·14 M-saline phosphate solvent was heated for 15 min. at 95°C. After adjusting the salt concentration to 0·6 M, 200 ml. of the solution was applied to a column prepared from 100 ml. of an MAK suspension. The DNA was then eluted with increasing salt increments collected in 5-ml. vol. Total recovery of absorbance at 260 mμ was 54 per cent. (After Roger, Beckmann and Hotchkiss (1966))

strands, and the results indicated that a separation between the two complementary strands had in fact been achieved. Furthermore, they showed that RNA could be synthesized enzymically on both strands, and that a very high degree of complementarity, with regard to base composition of the products, existed. Rudner, Karkas and Chargaff (1968b) also examined the two strands by direct base analysis and, in addition to confirming the complementarity, discovered that (A + C) was equal to (G + U). Seven bacterial species and the T_4 phage from *E. coli* were also investigated but the results were unexpected: (A + C) was equal to (G + U) for AT-rich DNA's, but not for DNA containing an equimolar proportion of GC and AT, nor the GC-rich types (Rudner, Karkas and Chargaff, 1969). The manner in which MAK columns enable the complementary strands

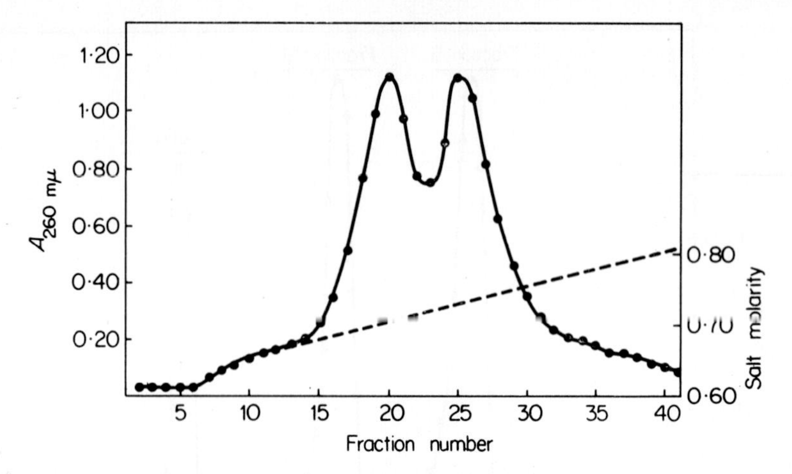

Figure 6.11. Elution of denatured pneumococcal DNA from MAK with a cold linear salt gradient. A total of 50 O.D. units of alkali-denatured DNA was applied to a 14 × 1·5 cm. MAK column in 0·60 M saline phosphate. Both the column and gradient mixing reservoir were jacketed and thermostated at 5–6°C throughout. Optical densities are indicated by the solid points. The eluting salt gradient was nearly linear between 0·60 and 0·80 M phosphate-buffered NaCl and is indicated by the dashed line. Fractions were collected in 3·25-cc vol. Recovery of optical density was 100 per cent. (After Roger (1968))

of DNA to be separated is not certain. However, Kubinski, Opera-Kubinska and Szybalski (1966) carried out experiments which seem to indicate that some DNA's possess local sequences of nucleotides which are rich in cystosine, and that these regions usually occur in only one of the complementary strands. Thus MAK may be responding only to these regions, but this is by no means certain.

RNA fractionation

The use of MAK chromatography during the investigation of RNA function and metabolism is very wide, and varied. Sueoka and Cheng (1962) first reported a separation of tRNA and the two rRNA components, but no fine structure was seen with the tRNA peak. Sueoka and Yamane (1962), however, specifically labelled tRNA molecules using individual (^{14}C) amino acids and fractionated the mixtures on MAK columns. They showed that within the 4S tRNA peak, definite regions existed for each tRNA molecule, and, out of the sixteen amino acids tested, eight consisted of more than one component. It was also possible to separate two individual

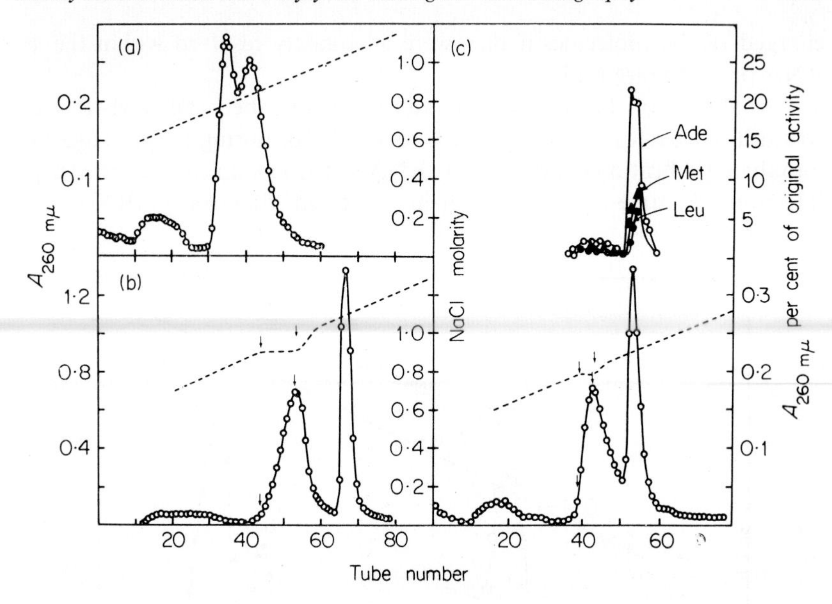

Figure 6.12. Elution of denatured *B. subtilis* DNA from MAK columns with the use of a linear salt gradient (a) and of an intermittent gradient elution technique (b and c).

(a) Heat-denatured DNA strain W23, eluted with a linear salt gradient between 0·6 and 1·2 M NaCl (total volume 400 ml.). Recovery (as per cent of input DNA): total, 64; fraction L,21; fraction H,31.

(b) Alkali-denatured DNA of strain W23, eluted with an intermittent salt gradient between 0·7 and 1·4 M NaCl (total volume, 450 ml.). As indicated by the arrows on the graph, the gradient was cut at tube 44 and reconnected at tube 53. Recovery (as per cent of input DNA): total, 80; fraction L, 42; fraction H, 31.

(c) Heat-denatured DNA of strain W23, eluted with an intermittent salt gradient between 0·6 and 1·2 M NaCl (total volume, 400 ml.). The gradient was cut at tube 39 and reconnected at tube 43. Recovery (as per cent of input DNA): total, 88; fraction L,42; fraction H, 40. The biological activity of individual fractions after renaturation at 68° for 3 h. is shown in upper portion of graph.

Transformation assays for adenine, leucine and methionine were performed after each fraction was adjusted to O.D.$_{260}$ of 0·1 and annealed at the eluting salt concentration. The recipient strain was Mu8u5u16. The residual activities of the denatured DNA before fractionation were (as per cent of original activity): adenine, 2·3; leucine, 1·1; methionine, 1·0. When the mixture of pooled fractions L and H was annealed, the respective activities were 34, 28, 31. (After Runder, Karkas and Chargaff (1968)) charged tRNA molecules if they were adequately resolved within the 4S RNA peak (Figure 6.13).

The ability of MAK columns to distinguish between acyl-tRNA's was utilized by Kano-Sueoka and Sueoka (1966) during an investigation into the modification of amino acyl-tRNA's of *E. coli* during bacteriophage infection. A time-course experiment revealed that leucyl-tRNA was

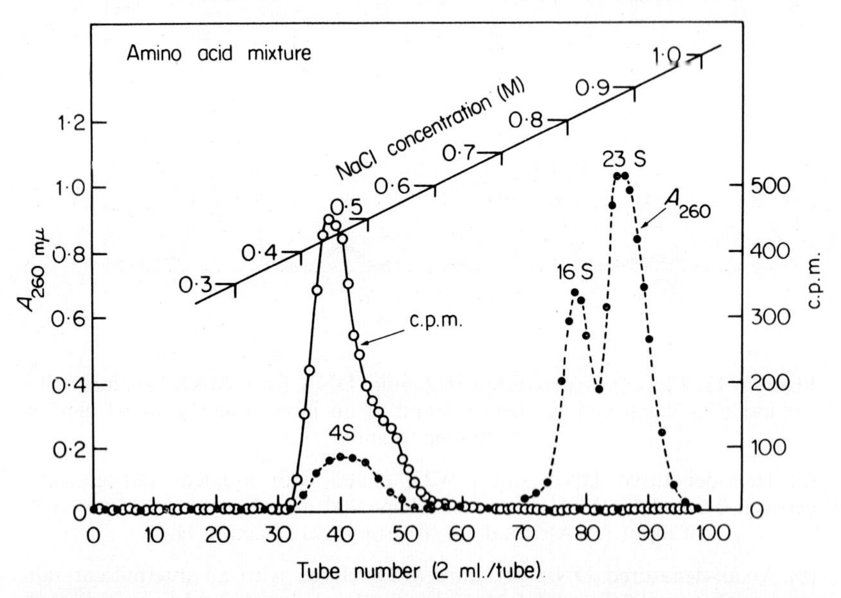

Figure 6.13. Elution pattern of ^{14}C-amino acid-labelled *E. coli* RNA on a methylated albumin column. A dialysed crude extract (15,000 g. supernatant) of logarithmically growing *E. coli* B harvested from a one-litre minimal medium culture was incubated with 100 μ curies of ^{14}C-yeast protein hydrolysate (50 μc./mg. Schwarz BioResearch Inc.). RNA was isolated by the phenol method, and 2·5 mg. of RNA was subjected to the column fractionation. The starting buffer was 0·2 M NaCl and the final buffer was 1·2 M NaCl. NaCl concentration was determined by measuring the refractive index of every tenth fraction. (After Sueoka and Yamane (1962))

altered very early after phage infection, and that the modification was related to the transition from the early to the late phase of phage protein synthesis. tRNA's can also be modified in other ways. Lowrie and Bergquist (1968) reported that tRNA, synthesized in the presence of 5-fluorouracil (Fu), contained up to 100 per cent replacement of uracil by Fu, and that the content of certain minor bases was also altered. Furthermore Fu-tRNA would accept amino acids to normal loading levels, and could transfer them to peptides in an *E. coli* polypeptide synthesizing system. No lysine could, however, be converted to polypeptide under the direction of bacteriophage R_{17} RNA after infection. The Fu-tRNA was separated from contaminating unsubstituted tRNA by MAK chromatography, and found to possess an altered secondary structure as judged by its thermal denaturation profile.

High molecular weight RNA's elute from MAK columns after DNA (Mandell and Hershey, 1960; Sueoka and Cheng, 1962), and can be resolved into two major components corresponding to the two rRNA species of the ribosomal sub-units. Otaka, Mitsui and Osawa (1962) isolated the ribonucleic acid synthesizing system from *E. coli* and studied the incorporation of 8-^{14}C-ATP into newly synthesized RNA. Salmine sulphate was used to protect the newly formed RNA from degradation and, after a 10 min. incubation, the total RNA and DNA was fractionated on MAK, and the fractions assayed for radioactivity. Four radioactive peaks were observed (Figure 6.14), one coincident with the DNA peak, one preceding the 16S ribosomal peak and having an S-value of 10–12, one before the 23S ribosomal peak and having an S-value of 17–19 and one post 23S ribosomal peak which contained the most label and had an

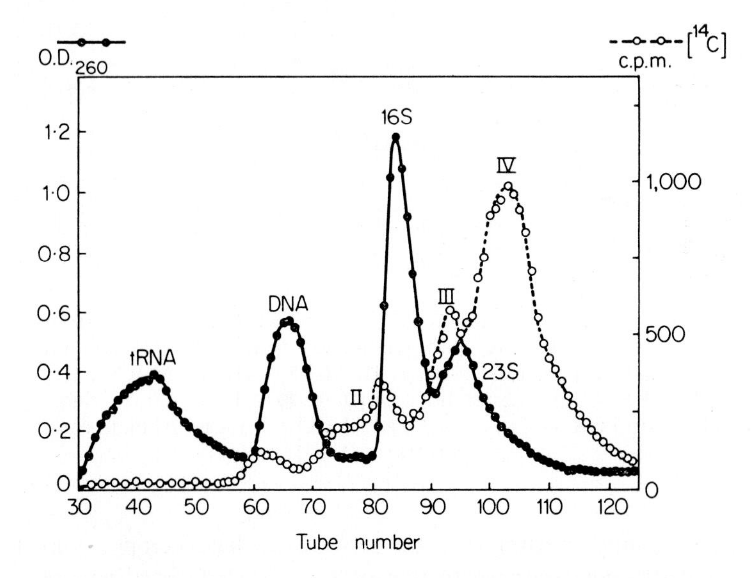

Figure 6.14. Fractionation of the RNA synthesized in the 105,000 x g supernatant of *E. coli* with methylated serum albumin column chromatography. 4 ml./tube/10 min. were collected at room temperature. (After Otaka, Mitsui and Osawa (1962))

S-value of the order of 26–30S and corresponded to the ribosomal RNA precursor.

Columns of MAK, which have been used successfully to fractionate rRNA, tRNA and DNA are also effective in separating species of mRNA (Ishihama, Mizuno, Takai, Otaka and Osawa, 1962; Monier, Naono, Hayes, Hayes and Gros, 1962). Ishihama and coworkers (1962) were able to fractionate total nucleic acid from *E. coli* after infection with T_2 bacteriophage. tRNA, *E. coli* DNA, T_2 phage DNA, 16S rRNA and 23S rRNA were eluted from MAK with NaCl molarities of 0·4, 0·6, 0·65, 0·75 and 0·85 M, respectively (Figure 6.15). Moreover, rapidly-labelled RNA was separated into four components, one of which could

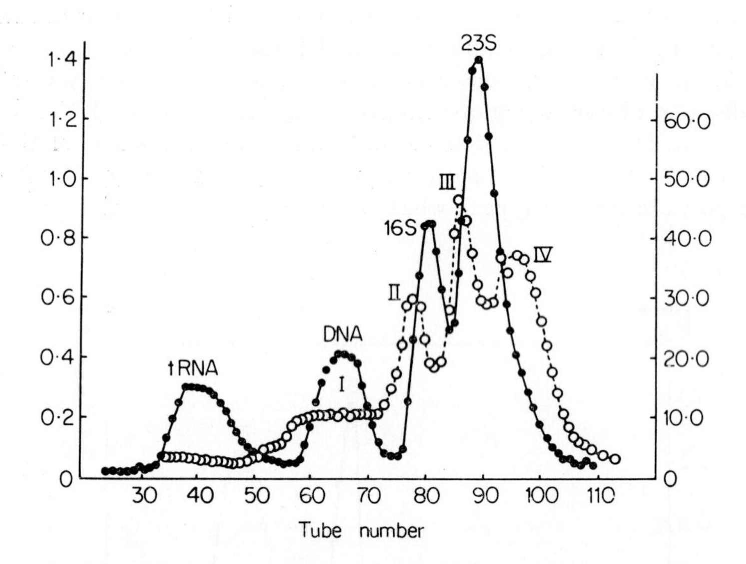

Figure 6.15. Fractionation of ^{32}P-labelled nucleic acids from growing *E. coli* cells, with methylated serum albumin column chromatography. I, M-RNA-I; II, M-RNA-II; III, M-RNA-III; IV, M-RNA-IV. Left-hand ordinate —●—, O.D.$_{260}$/ml.; right-hand ordinate, ---○---, c.t.s./min./ml. (X10^{-3}). (After Ishihama and coworkers (1962))

have the affinity to complex with DNA. The base composition of all four fractions corresponded to that of *E. coli* DNA. Furthermore, after infection with T_2 bacteriophage, three of the fractions had a base composition similar to that of T_2 DNA. Fraction IV associated specifically with the 50S and 30S rRNA but not with the 70S, and its synthesis was inhibited by chloramphenicol. Fractions II and III did not combine with rRNA. Only fraction IV was postulated to represent the true 'messenger RNA' synthesized in *E. coli* cells after infection with T_2 bacteriophage (Ishihama and coworkers, 1962).

Monier and his coworkers (1962) were able to separate 4S, 16S and 23S RNA's when total nucleic acid was isolated from *E. coli* and fractionated on a MAK column (1·2 × 30cm.). Moreover, they separated the rapidly-labelled RNA into three fractions which were termed α, β and γ (Figure 6.16), and which had different sedimentation values (23S to 30S). All three fractions were found to stimulate the incorporation of ^{14}C-amino acids into sub-cellular systems.

The virus Ø-X174 contains only single-stranded DNA (Sinsheimer, 1959) and when *E. coli* cultures are infected with the virus, the complement to the injected strand is immediately synthesized, and a double-stranded DNA molecule results, which has been termed 'replicating form' (RF-DNA) (Sinsheimer and coworkers, 1962). The RF-DNA has been isolated and extensively purified on MAK by Hayashi, Hayashi and Spiegelman (1963) who also studied the specific Ø-X174 messenger RNA synthesized after infection. Cultures of *E. coli* were pulse-labelled with ^{3}H-uridine for 90 sec., after 50 min. infection, and total RNA was isolated and fractionated using MAK (Figure 6.17). Five fractions were selected and their capacity to hybridize with RF-DNA tested. Each fraction was individually added to RF-DNA (which had been heated to separate the complementary strands), and the resulting mixture incubated with ribonuclease to remove RNA, and subsequently fractionated on MAK. A peak of RNA hybridizable to RF-DNA was observed in the region corresponding to fraction IV (Figure 6.17), the position at which denatured viral DNA also eluted. In order to confirm these results, both single and RF-DNA of the virus Ø-X174 were labelled with ^{32}P and the RNA with ^{3}H. When vegetative single-stranded Ø-X174 DNA (heated or not) was used to hybridize with ^{3}H-labelled RNA and chromatographed on a MAK column no hybrid was formed (Figure 6.18, a and b). On the other hand, when heat-denatured ^{32}P RF-DNA was used to hybridize with ^{3}H-RNA followed by fractionation on MAK, a fraction was obtained which contained both ^{3}H and ^{32}P (Figure 6.18c) and was resistant to ribonuclease digestion. All these findings led Spiegelman and his coworkers to conclude that Ø-X174 messenger RNA is incapable of complexing with single-stranded DNA, but can hybridize with RF-DNA which contains both complementary strands.

Kubinski and Koch (1962, 1963) used MAK to separate infective RNA from cells infected with poliovirus. Ribosomal RNA was eluted from the column before the infective RNA.

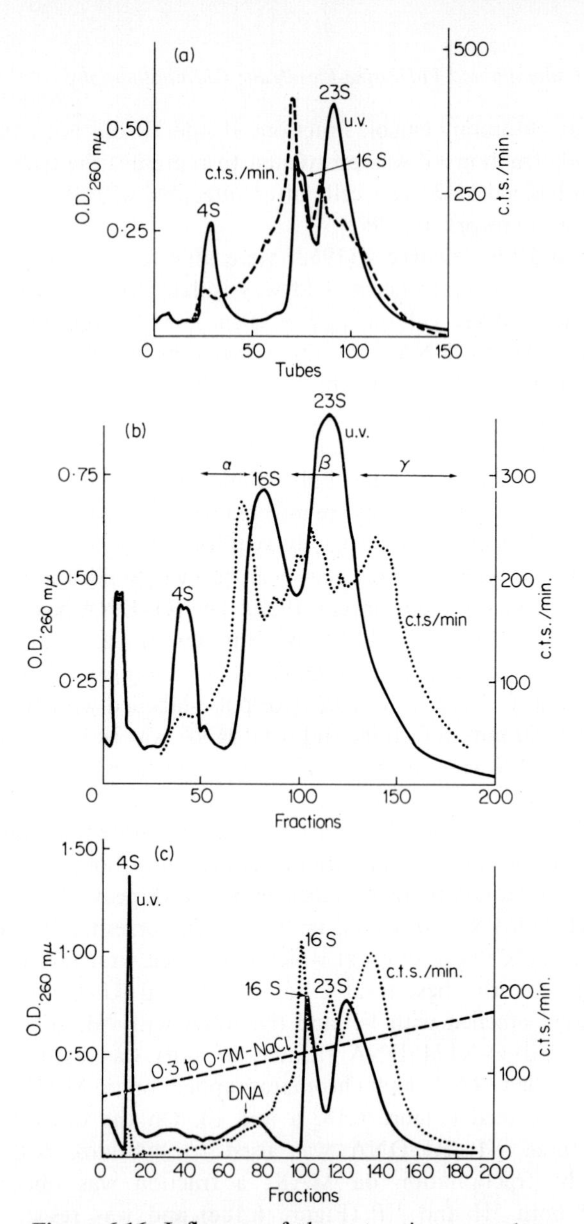

Figure 6.16. Influence of the extraction procedure upon the chromatographic behaviour of the pulse-labelled RNA fraction from *E. coli*. Cells from the strain ML 308 were labelled with ^{14}C uracil, and the exponential elution gradient technique of Mandell and Hershey was used. (a) RNA prepared by grinding cells with alumina. (b) RNA obtained following lysis with SDS. (c) Cell disruption is obtained by the use of a French Press. (After Monier and coworkers (1962))

MAK chromatography has not been confined to the study of bacterial systems. Roberts and D'Ari (1968) applied zonal centrifugation and MAK chromatography to the problem of base sequencing ribosomal and 'ribosomal precursor' RNA from Ehrlich Ascites cells, while Ellem (1966) has investigated mammalian DNA-like RNA. Two forms of DNA-like RNA could be detected; one being eluted by a normal salt gradient and having a high S-value (50S), and the other requiring a more vigorous elution and having a lower modal S-value (16S). Both types were rapidly

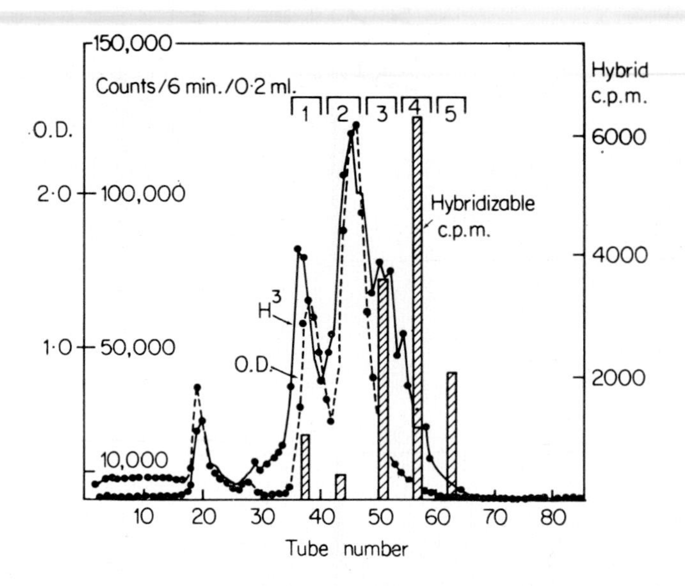

Figure 6.17. Column identification of Ø-X174 specific RNA. Ø-X174 infected *E. coli* cluture was pulse-labelled with ^{3}H uridine for 90 sec., 50 min. after infection. The total RNA was isolated and chromatographed. O.D. profile identifies preexisting RNA. The pooled samples in the regions indicated in the figure were concentrated. The same number of counts from each sample was hybridized with 20 μg. of RF-DNA which had been heat-denatured in 1/10 SSC (0·15 M NaCl, 0·015 Na citrate) at 97–98°C for 15 min. Hybridization was carried out in 2 × SSC at 42·5°C for 18h. The reaction mixture was chilled and 30 μg./ml. of pancreatic RNase, free of contaminating DNase was added. RNase treatment was performed at 26°C for 30 min. The reaction mixture was then loaded on an MAK column. Counts in the hybrid region were summed up and are shown in the bar-histograms. (After Hayashi, Hayashi and Spiegelman (1963))

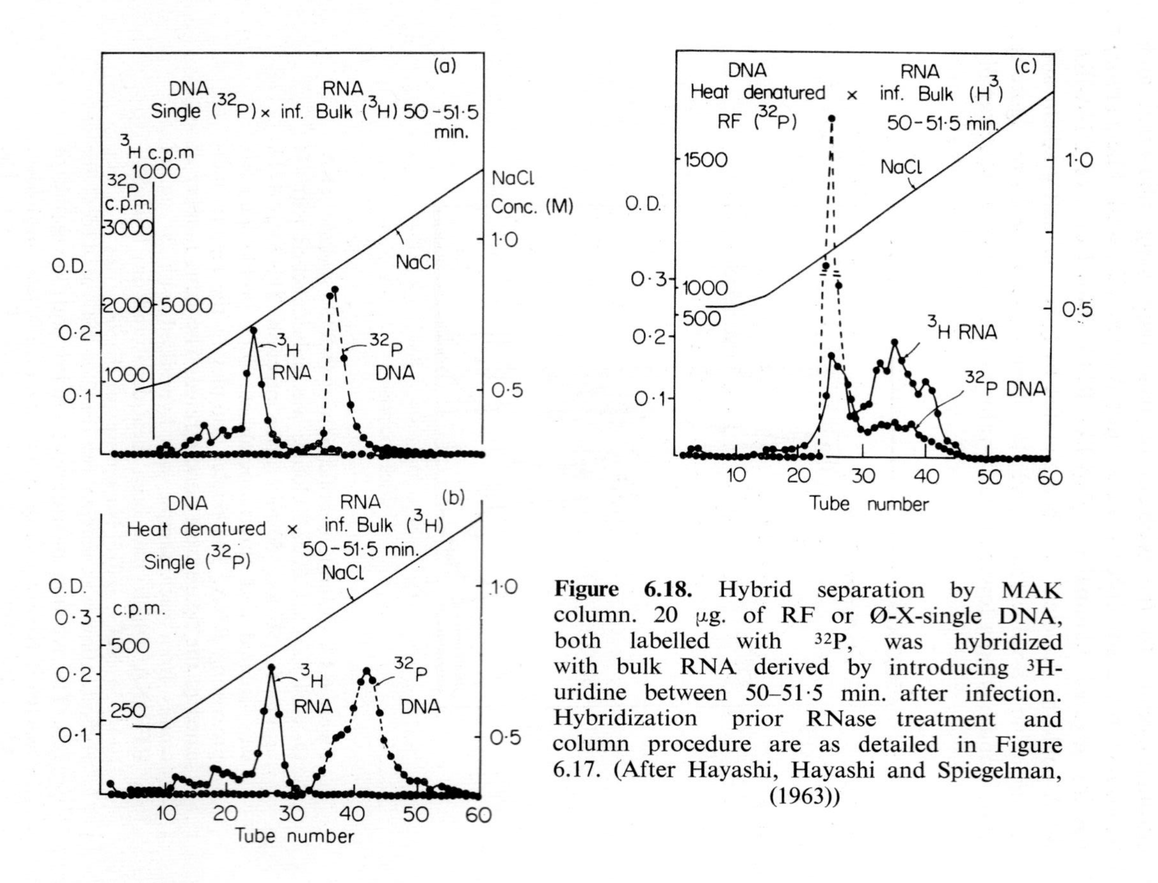

Figure 6.18. Hybrid separation by MAK column. 20 μg. of RF or Ø-X-single DNA, both labelled with ^{32}P, was hybridized with bulk RNA derived by introducing ^{3}H-uridine between 50–51·5 min. after infection. Hybridization prior RNase treatment and column procedure are as detailed in Figure 6.17. (After Hayashi, Hayashi and Spiegelman, (1963))

synthesized, and the 16S-type resembled messenger RNA in several of its properties. Refinements in the resolution of this technique were introduced by Ellem (1967) using a dual label to compare the rates of synthesis of nucleic acid fractions, and the method used to confirm the selective inhibition of ribosomal RNA in HeLa cells by low doses of actinomycin D.

Early studies on RNA synthesis in *E. coli* infected with T-even bacteriophage showed that net synthesis of RNA ceased immediately following infection (Cohen, 1947). However, Baguley, Bergquist and Ralph (1967) were able to detect a low molecular weight, radioactive RNA fraction by MAK chromatography in *E. coli* following infection with T_4 phage. This radioactive RNA hybridized with T_4 phage DNA but not with *E. coli* DNA. These and other studies (Bautz, 1963) have demonstrated the synthesis of a mRNA which is phage-specific, and an additional low molecular weight RNA, which the authors consider may be functioning in a similar manner to 5S RNA in other systems.

Okamoto and Kawade (1963) suggested that silicic acid could replace kieselguhr as a support for methylated-albumin and this idea was applied by Stern and Littauer (1968). They used a methylated-albumin-silicic acid column to fractionate tRNA from *E. coli*, and claimed that the column had a 100-fold greater adsorptive capacity, together with a higher resolution for tRNA, than a similar MAK column.

(B) Polylysine-kieselguhr (PLK)

PLK preparation

Kieselguhr, the diatomaceous earth used for MAK chromatography, is also the supporting matrix for PLK columns. Impurities and u.v. absorbing material are first removed by washing the kieselguhr with distilled water and the 'fines' discarded by allowing the kieselguhr to settle and decanting the cloudy supernatant. The washed material is then dried in an oven before storing. Polylysine, is supplied commercially, and stock solutions of 10mg. per ml. of 0·4 M NaCl containing 0·02 M KH_2PO_4 (0·4 M buffered saline) pH 6·7 are prepared and stored at 4°C.

A polylysine slurry is formed by suspending the required amount of kieselguhr in 0·4 M buffered saline (5ml./g. kieselguhr) and stirring the suspension to an even consistency. The suspension is heated until it begins to boil and immediately cooled to remove the air trapped within the kieselguhr particles. The boiled kieselguhr is resuspended in 0·4 M buffered saline, and the required amount of polylysine solution (1mg. polylysine/g. kieselguhr) pipetted into the suspension with constant stirring. The resulting PLK suspension forms the fractionating layer of the column.

Figure 6.19 shows a completed PLK column prepared by the method of Ayad and Blamire (1968). PLK columns are eluted with linear gradients produced by the apparatus shown in Figure 6.20 devised by Ayad, Bonsall and Hunt (1968). The following shorthand notation has been adopted to describe a particular linear gradient: *X*ml. : 0·4 to 4 M, where *X* is the initial volume of 0·4 M NaCl, and 0·4 to 4·0 M is the range of salt molarity used (buffered to pH 6·7 with 0·02 M KH_2PO_4).

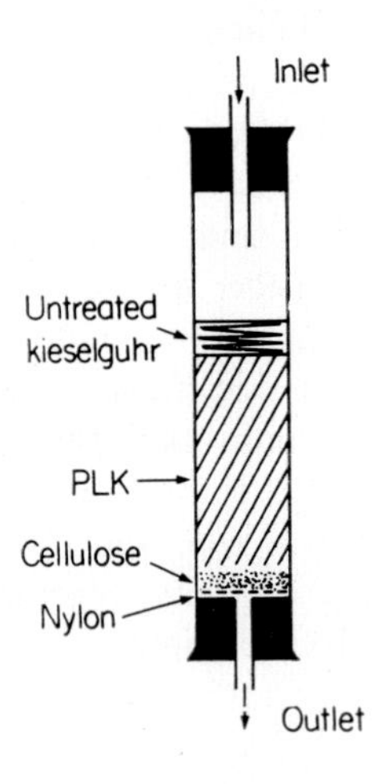

Figure 6.19. Schematic representation of polylysine-kieselguhr column (PLK). (After Ayad (1966))

Fractionation of nucleic acids

DNA from *B. subtilis* (100μg./ml. 0·4 M buffered NaCl) prepared by the method of Marmur (1961), was applied to a PLK column (10g.) using air pressure (2lb./sq.in.) and eluted with 150ml. : 0·4 to 4·0 M gradient (Ayad and Blamire, 1968). The resulting elution profile is shown in Figure 6.21. Three peaks of u.v. absorbing material could be observed, each giving a positive reaction with the diphenylamine assay for DNA (Burton, 1956). The first peak, eluting at 0·6 M NaCl, was variable in size, and from analytical data was concluded to be low molecular weight material. The second, central peak at 1·2 M NaCl was always low and broad, and the small leading shoulder of this peak indicated some heterogeneity. The analytical data were indicative of diffuse material of no definite configuration. The final peak, eluting at 1·95–2·0 M NaCl, gave analytical figures corresponding to high molecular weight, native DNA molecules.

Subsidiary experiments demonstrated that the fractionation was genuine and not an artifact of the column or elution method. Fractions from each of the three peaks were pooled separately and the salt molarity re-adjusted to 0·4 M NaCl. Each individual peak was then rechromatographed on different columns in an identical manner to that described previously.

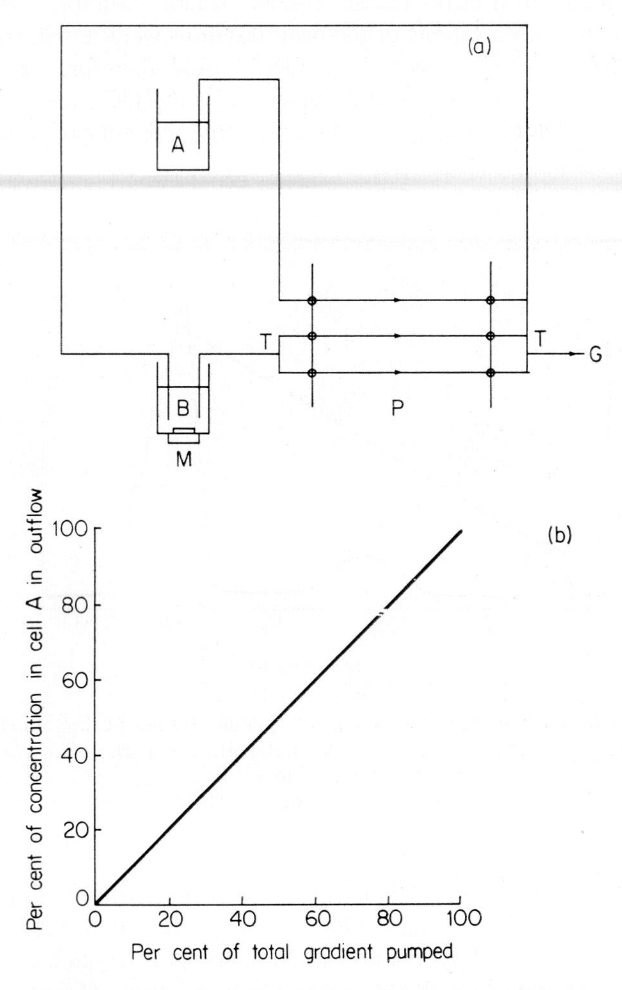

Figure 6.20. (a) Arrangement of apparatus for production of linear gradients: (A) beaker containing solution at high concentration, (B) beaker containing solution at low concentration, (M) magnetic stirrer, (P) peristaltic pump with three channels, (T) T-joints, (G) outflow of gradient. (b) Plot of percentage of total volume of gradient pumped versus percentage of concentration in beaker A, in the outflow (see text). (After Ayad, Bonsall and Hunt (1968))

In each case, the profile and salt molarity were almost identical to those found originally, and no other peaks appeared. When the linear gradient was changed from 150ml. : 0·4 to 4·0 M to 150ml. : 0·4 to 3·0 M, the second and the third peaks eluted at higher fraction numbers. However, when the salt concentrations of the peak fractions were tested, they were identical to the original molarities. DNA could therefore be retained by PLK columns and subsequently be eluted with NaCl, the position of elution being determined by the NaCl molarity and not the column size or eluting buffer.

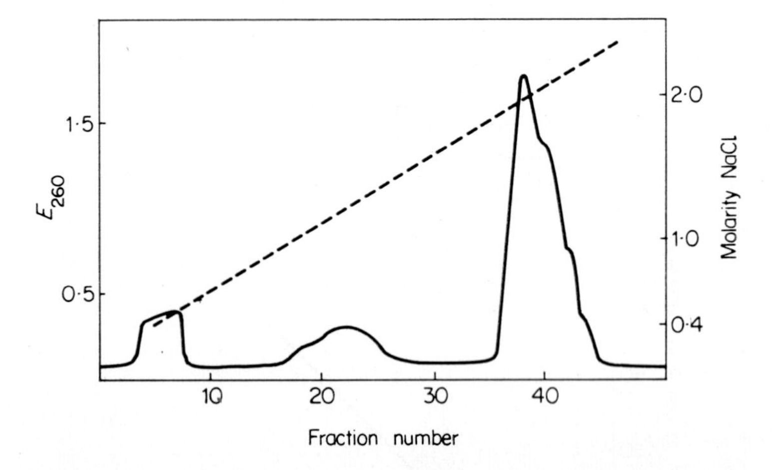

Figure 6.21. The continuous elution profile (O.D. at 260 mμ) of effluents from PLK column (—) loaded with native *B. subtilis* DNA (1·5 mg. in 15 ml.) and eluted with linear gradient of NaCl (see text). (After Ayad and Blamire (1968))

The three peaks were also assayed for base composition and found to possess different GC contents. Base composition of DNA therefore influences its mode of fractionation on PLK columns (Ayad and Blamire, 1968). To investigate whether this was the only source of fractionation, a sample of DNA was sonicated to reduce its molecular weight, and a second sample heat-denatured to separate the DNA strands. When fractionated on PLK, both samples gave similar elution profiles to that of native DNA. Sonicated DNA eluted in three regions at 0·6, 1·25 and 1·84 M NaCl, while heat-denatured eluted at 0·6, 1·35 and 2·07 M NaCl. The main factor influencing the elution of DNA on PLK is therefore base composition, since changes in molecular weight and secondary structure do not significantly affect the elution profile. These findings were further confirmed by fractionating calf thymus, *B. subtilis* and *E. coli* DNA's on PLK (Ayad and Blamire, 1969). The three DNA's differ in GC content and the main peak of each DNA species eluted at a different salt molarity.

Since PLK separates DNA regions which differ in chemical structure, the fractionation of individual biologically active markers was tested (Ayad and Blamire, 1969). Successive fractions across the DNA peak, eluted with 2 M NaCl during the fractionation of total nucleic acids from wild-type *B. subtilis*, were tested for their ability to transform the mutant strains of *B. subtilis* which were deficient either in the tryptophan gene or in both the tryptophan and histidine genes. In each case, the biological activity for the genes was not uniformly distributed within the 2 M peak, but was a maximum in certain fractions, indicating a certain degree of gene separation.

The fractionation of RNA on PLK was first studied using total nucleic acid mixtures isolated from *B. subtilis* (Ayad and Blamire, 1969). A typical elution pattern of a mixture which had been deproteinized twice with chloroform/isoamyl alcohol is shown in Figure 6.22. A PLK column (5g.) was used and the gradient was 100ml. : 0·4 to 4·0 M. Four major peaks of u.v. absorbing material were observed with maxima at 0·4 to 0·5, 1·0 to 1·1, 1·5 to 1·6 and 1·95 to 2·0 M NaCl, respectively. In certain preparations, a region could be observed prior to the 1·0 to 1·1 M NaCl peak, eluting in the range of 0·65 to 0·9 M NaCl. The material eluting in

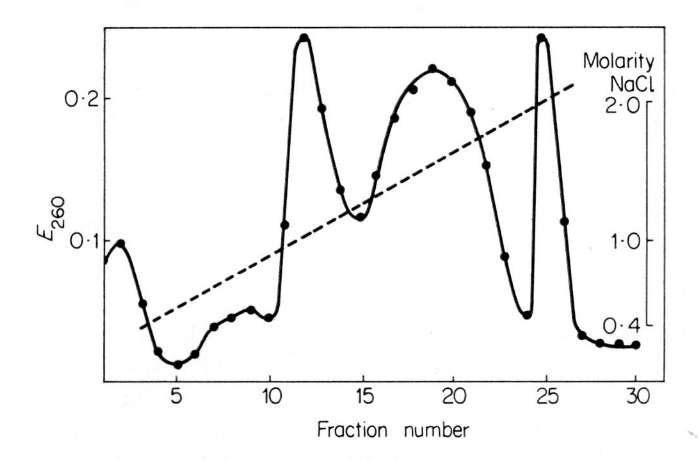

Figure 6.22. Elution profile at 260 mμ of a nucleic acid mixture isolated from *B. subtilis* and fractionated on 5 g. PLK column using NaCl linear gradient between 0·4 and 4·0 M (see text). (After Ayad and Blamire (1969))

the various peaks was identified by fractionating mixtures treated with either RNase or DNase and by reference to the elution patterns of known standards (Ayad and Blamire, 1969). DNase treatment removed the final 1·95–2·0 M peak, and RNase treatment removed the 1·0 to 1·1 M and 1·5 to 1·6 M peaks. In both cases, there was an increase in the amount of material eluting in the first 0·4 to 0·5 M peak which was assigned to low molecular weight oligonucleotides. The 1·95 to 2·0 M peak was DNA and the 1·0 to 1·1 M and 1·5 to 1·6 M peaks were RNA species and this was confirmed by chemical assay for RNA and DNA. Protein assays of the individual fractions showed that most of the protein was not retained by the column to any significant extent and eluted with the washing buffer or early in the elution profile.

Fractionation of various RNA standards identified the RNA species eluting at the various salt molarities (Ayad and Blamire, 1969, 1970). The elution profiles are shown in Figures 6.23, 6.24, 6.25 and 6.26. The

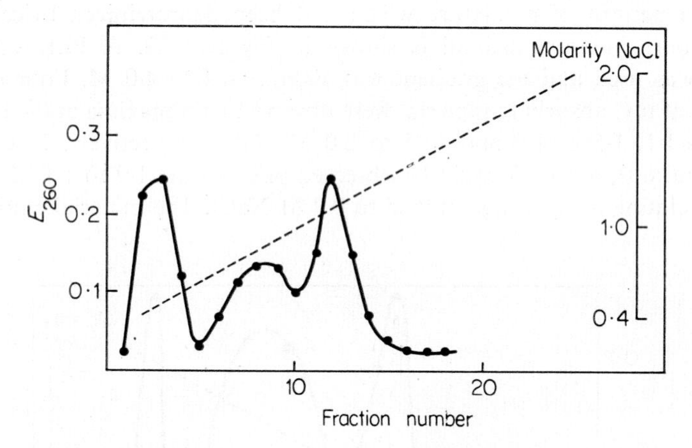

Figure 6.23. Fractionation of tRNA type XI (Sigma, Chem. Co.) on 2·5 g. PLK column using linear gradient of NaCl (see text). (After Ayad and Blamire (1970))

main tRNA (from yeast) peak eluted at about 1·1 M NaCl and was similar to that found in nucleic acid mixtures. There was also a considerable amount of material in the peak eluting prior to the 1·1 M peak, in addition to the presence of low molecular weight oligonucleotides. A commercial ribonucleic acid 'core' (resistant to RNase digestion) fraction gave a broad peak at 0·80 to 1·35 M NaCl, which was preceded by two smaller peaks at lower salt molarities. The resistance of the material from the 0·8 to 1·35 M NaCl region to RNase digestion was tested by subjecting it to 50μg./ml. RNase at 37°C for 30 min. The digest fractionated as before, thus confirming the resistance of 'core' RNA to RNase. Highly polymerized RNA from *E. coli* fractionated as a single peak with a maximum at 1·85 to 1·95 M NaCl. The peak commenced eluting at 1·15 to 1·2 M NaCl and gave a characteristic long leading edge. A replicative double-stranded RNA from R_{17} phage also gave a single peak very similar to that obtained with highly polymerized RNA.

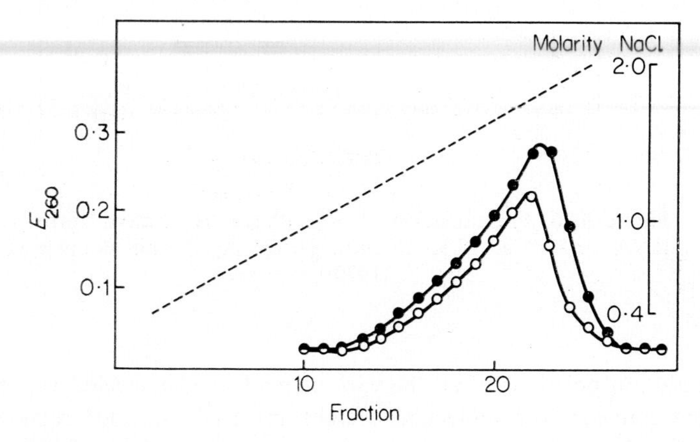

Figure 6.24. Fractionation of highly polymerized RNA on 2·5 g. PLK column, ○—○, standard RNA (BDH, Ltd.) and ●—●, prepared from *E. coli* strain MRE 600 (After Ayad and Blamire (1970))

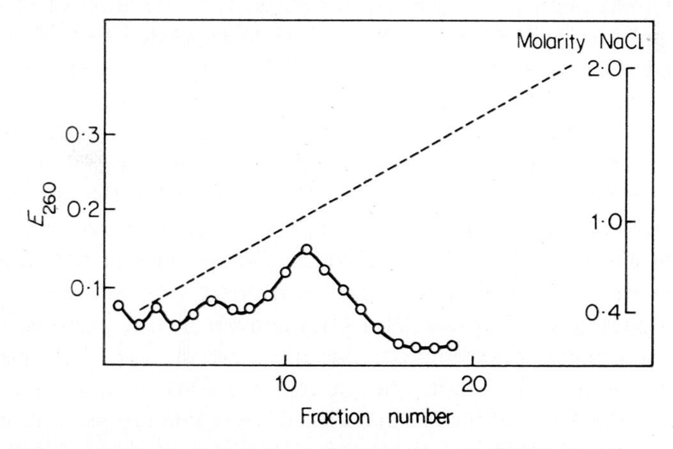

Figure 6.25. Fractionation of RNA 'core' (BDH, Ltd.) on 2·5 g. PLK column. (After Ayad and Blamire (1970))

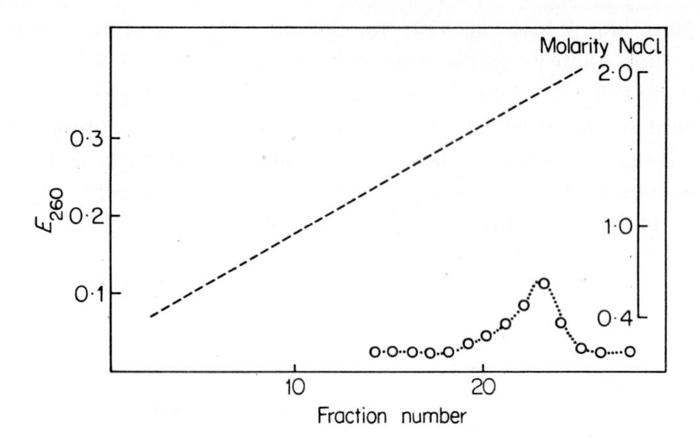

Figure 6.26. Fractionation of R_{17} phage 'replicative' form RNA on 2·5 g. PLK column. (After Ayad and Blamire (1970))

The elution positions of all the various nucleic acid species were therefore determined: low molecular weight material was not retained or easily eluted on increasing the salt molarity above 0·4 M, tRNA eluted as a sharp peak at 1·1 M NaCl, DNA (native, high molecular weight) as a sharp peak at 1·95 to 2·0 M and highly polymerized rRNA eluted in the region between tRNA and DNA.

The heterogeneous nature of the RNA peak eluting between 1·25 and 1·75 M (maximum 1·5 to 1·6 M) compared to the elution of standard highly polymerized rRNA at 1·85 to 1·95 M suggested that rRNA was degraded during isolation. Various modifications were therefore made which resulted in the preparation of highly polymerized rRNA in the nucleic acid mixture (Ayad and Blamire, 1970). The fractionation of this nucleic acid mixture (after DNase treatment) is shown in Figure 6.27. There were two small peaks of material which eluted at the salt molarity where tRNA elutes, but the major components eluted as two overlapping peaks at about 1·7 M and 1·8 M NaCl. The S-values of the individual fractions were estimated, and it was found that the 1·7 and 1·8 M peaks corresponded to the 16S and 23S rRNA components, respectively.

The main factor governing the elution of the various RNA species is molecular weight. However, the elution positions of double-stranded R_{17} phage RNA and of the 16S and 23S rRNA would suggest that secondary structure and conformation may also be important (Ayad and Blamire, 1970).

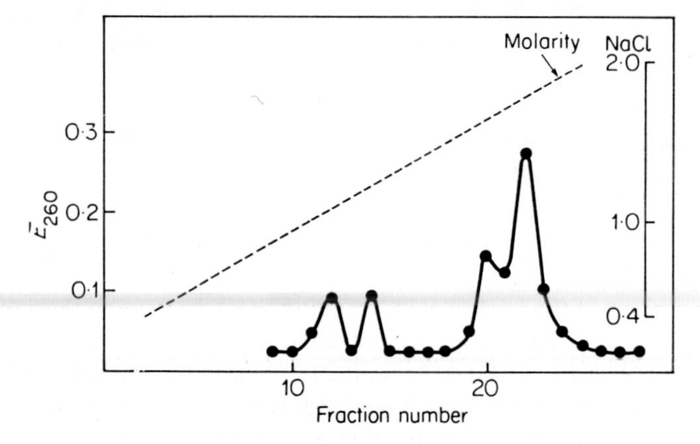

Figure 6.27. Fractionation of ribosomal RNA from *B. subtilis* on PLK column using linear gradient of NaCl. (After Ayad and Blamire (1970))

PLK chromatography has also been used to study the more specific problems associated with nucleic acid metabolism. Kemp (1968) and Detlfsen (1970) studied the DNA–RNA hybrids in *Pisum arvense*, and the separation of plasmid DNA from chromosomal DNA in the drug-resistant strain of *S. aureus* was effected by Cannon and Dunican (1970). The method is also finding application in the investigation of RNA virus infection of chick embryo cells (Bromley and Barry, 1970). The fractionation of a rapidly-labelled nucleic acid from *Rhodospirillum rubrum* grown anaerobically in the light and aerobically in the dark may help to elucidate the different biosynthetic processes which occur during heterotrophic and photosynthetic growth (Ayad, Premachandra and Gregory, 1971).

The characteristic eluting position of all the major nucleic acid species from PLK columns allows the identification and quantitative estimation of any species in a nucleic acid mixture. It is also possible to assess the degree of purity and homogeneity of each species since any variation in the structural integrity will be reflected in the elution profile.

Summary

MAK and PLK chromatography are similar in that both systems use kieselguhr as the supporting matrix for the fractionating substance (protein or polypeptide), and the nucleic acids are eluted using increasing concentrations of NaCl.

The mechanism of MAK fractionation appears to be far more complex than that of PLK chromatography and is not completely understood. The relative importance of each of the three main factors (molecular weight, secondary structure and base composition) governing the elution from MAK is not known, and in this respect the elution of DNA after tRNA but prior to rRNA is highly anomalous. The basis of PLK fractionation has been fairly well characterized. It depends on the polycation–nucleic acid interactions described in Chapter 6. Investigations using dye-binding studies and competitive exchange experiments indicate that certain regions of the nucleic acids (binding sites), because of their secondary structure and base composition, have a particular affinity for short lengths of polylysine, and it is the number of binding sites which determines the salt molarity of elution (Ayad and Blamire, 1970). These specific interactions allow DNA to be fractionated according to its base composition, the secondary structure and molecular weight having little affect on the elution profile. RNA on the other hand is fractionated mainly according to its molecular weight, the degree of secondary and tertiary structure perhaps influencing the elution profile to a certain, but not significant, extent.

Many factors (in addition to those discussed above) contribute to the irreproducibility of MAK chromatography. In particular, the elution profile obtained depends to a large extent on the degree of methylation of serum albumin, a factor which cannot easily be controlled. Moreover, the methylated-albumin is a much more complex and heterogeneous molecule than polylysine, and it is therefore more difficult to assess the factors influencing fractionation. A further factor influencing the reproducibility of fractionation is the amount of material which may be recovered from a column after chromatography. Only about 60 per cent native DNA and about 20 per cent denatured DNA can be recovered from MAK.

It is therefore not possible to determine the nature of the material which adheres to the column, nor to be certain as to the nature of the material actually eluted. On the other hand, the recovery of native and denatured DNA from PLK columns is 100 per cent and 70 per cent, respectively.

PLK chromatography has many other advantages over MAK chromatography. There is a tendency to 'channel' in the case of MAK, whereas no channelling has been observed using PLK even when large amounts of materials have been loaded (50mg./10g. PLK column). Although both MAK and PLK chromatography allow the resolution of all the nucleic acid species into the main categories, DNA, tRNA and rRNA, neither method is effective in separating the various amino acid specific tRNA's within the tRNA profile. In this respect, MAK and PLK fractionation are superseded by ion-exchange chromatography and CCD.

PLK fractionation, being highly reproducible, quantitative and specific for DNA base composition, provides an efficient method for the separation of R-factor DNA (or plasmids in general) from chromosomal DNA. For example, transmission of the R-factor of *E. coli* (50 to 52 per cent GC) into *Proteus* (38 per cent GC) by conjugation, followed by extraction and PLK fractionation of the total native DNA, would result in the separation of the R-factor DNA from the chromosomal DNA of *Proteus* by virtue of their differing GC contents. Moreover, large amounts (2 to 5mg.) could be isolated by this method. The alternative fractionation procedures: nitrocellulose, CsCl and hydroxyapatite all require selective denaturation prior to fractionation and are less quantitative procedures. Similarly viral DNA and host DNA (with different GC content) can easily be separated in a single step by PLK chromatography.

Finally, studies on the interaction between PLK and DNA can help in the understanding of the way in which histones (lysine- and arginine-rich polypeptides) complex with DNA and control its function in mammalian cells.

References

Agarwal, K. L., H. Buchi, M. H. Caruthers, N. Gupta, H. G. Khorana, K. Kleppe, A. Kumar, E. Ohtsuka, W. L. Rajbhandary, J. H. Van De Sande, V. Sgaramella, H. Weber and T. Yamada (1970). Total synthesis of the gene for an alanine transfer ribonucleic acid from yeast. *Nature*, **227,** 27.

Ageno, M., E. Dore, C. Frontali, M. Arca', L. Frontali and G. Tecce (1966). Interaction between denatured DNA and RNA from *Bacillus stearothermophilus* involving only one-half of total DNA. *J. Mol. Biol.*, **15,** 555.

Ahonen, J., and E. Kulonen (1966). Separation of transfer ribonucleic acid from deoxyribonucleic acid by gel filtration. *J. Chromatog.*, **24,** 197.

Akrigg, A., and S. R. Ayad (1969). Competence-inducing factor of *Bacillus subtilis*. *Biochem. J.*, **112,** 13P.

Akrigg, A., and S. R. Ayad (1970). Studies on the competence inducing factor of *Bacillus subtilis*. *Biochem. J.*, **117,** 397.

Akrigg, A., S. R. Ayad and G. R. Barker (1967). The nature of competence inducing factor in *Bacillus subtilis*. *Biochem. Biophys. Res. Commun.*, **28,** 1062.

Akrigg, A., S. R. Ayad and J. Blamire (1969). Uptake of DNA by competent bacteria: A possible mechanism. *J. Theoret. Biol.*, **24,** 266.

Alberts, B. M. (1967). Efficient separation of single-stranded and double-stranded deoxyribonucleic acid in a dextran-polyethylene glycol two phase system. *Biochemistry*, **6,** 2527.

Albertsson, P. A. (1960). In *Partition of Cell Particles and Macromolecules*, John Wiley and Sons, New York.

Albertsson, P. A. (1965a). Partition studies of nucleic acids. Influence of electrolytes, polymer concentration and nucleic acid conformation on the partition in the dextran-polyethylene glycol system. *Biochem. Biophys. Acta.*, **103,** 1.

Albertsson, P. A. (1965b). Thin-layer countercurrent distribution. *Anal. Biochem.*, **11,** 121.

Albertsson, P. A., Y. Hanzon and G. Toschi (1959). Isolation of ribonucleoprotein particle from rat brain microsomes by a liquid two-phase system. *J. Ultrastruct. Res.*, **2,** 366.

Anacker, W. F., and V. Stoy (1958). Protein chromatography on calcium phosphate. Purification of nitrate-reductase from wheat leaves. *Biochem. Z.*, **330,** 141.

Anagnostopoulos, C., and J. Spizizen (1961). Requirements for transformation in *Bacillus subtilis*. *J. Bacteriol.*, **81,** 741.

Ando, T., and K. Suzuki (1966). The amino acid sequence of the second component of clupeine. *Biochim. Biophys. Acta*, **121,** 427.

Apgar, J., R. W. Holley and S. H. Merrill (1961). Countercurrent distribution of yeast "soluble ribonucleic acids" in a modification of the Kirby system. *Biochim. Biophys. Acta.*, **53,** 220.

Apgar, J., R. W. Holley and S. H. Merrill (1962). Purification of the alanine-, valine-, histidine-, and tyrosine- acceptor ribonucleic acids from yeast. *J. Biol. Chem.*, **237,** 796.

Araki, S. (1956). Structure of the agarose constituent of agar-agar. *Bull. Chem. Soc., Japan*, **29,** 543.

Archer, L. J., and O. E. Landman (1969a). Development of competence in thymine-starved *B. subtilis* with chromosomes arrested at the terminus. *J. Bacteriol.*, **97,** 166.

Archer, L. J., and O. E. Landman (1969b). Transport of donor DNA into the cell interior of thymine-starved *B. subtilis* with chromosomes arrested at the terminus. *J. Bacteriol.*, **97,** 174.

Arnott, S., M. H. F. Wilkins, W. Fuller and R. Langridge (1967). Molecular and crystal structures of double-helical RNA. An 11-fold molecular model and comparison of the agreement between the observed and calculated three-dimensional diffraction data for 10- and 11- fold models. *J. Mol. Biol.*, **27,** 535.

Arnstein, H. R. V., R. A. Cox and J. A. Hunt (1964). The function of high-molecular weight RNA from rabbit reticulocytes in haemoglobin biosynthesis. *Biochem. J.*, **92,** 648.

Aronson, A. I. (1962). Sequence differences between RNA's isolated from 30S and 50S ribosomes. *J. Mol. Biol.*, **5,** 453.

Attardi, G., R. C. Huang and S. Kabat (1965). Recognition of ribosomal RNA sites in DNA. *Proc. Nat. Acad. Sci., USA*, **54,** 185.

Avery, O. T., C. M. MacLeod and M. McCarty (1944). Studies on the chemical nature of the substance inducing transformation of pneumococcal type. *J. Exptl. Med.*, **79,** 137.

Ayad, S. R. (1964). Unpublished results.

Ayad, S. R. (1966). Polylysine kieselguhr column; devised after experimentation.

Ayad, S. R. (1968). Unpublished results.

Ayad, S. R. (1969). Inhibition of *Bacillus subtilis* transforming system by acriflavin. *FEBS Letters*, **2,** 348.

Ayad, S. R. (1970). Effect of ethidium bromide on DNA uptake and integration in *Bacillus subtilis* transformation system. Manuscript in preparation.

Ayad, S. R., and G. R. Barker (1969). The integration of donor and recipient deoxyribonucleic acid during transformation of *Bacillus subtilis. Biochem. J.*, **113,** 167.

Ayad, S. R., and J. Blamire (1968). Fractionation of *Bacillus subtilis* DNA by use of poly L-lysine kieselguhr columns. *Biochem. Biophys. Res. Commun.*, **30,** 207.

Ayad, S. R., and J. Blamire (1969). Fractionation of nucleic acids using poly-lysine kieselguhr chromatography. *J. Chromatog.*, **42,** 248.

Ayad, S. R., and J. Blamire (1970). Fractionation of ribonucleic acid and binding studies on columns of polylysine kieselguhr. *J. Chromatog.*, **48,** 456.

Ayad, S. R., and M. Fox (1968). DNA uptake by a mutant strain of lymphoma cells. *Nature*, **220,** 35.

Ayad, S. R., and M. Fox (1969). The implication of repair processes in the mechanism of DNA integration by lymphoma cells. *Intern. J. Radiation. Biol.*, **15,** 445.

Ayad, S. R., and M. Fox (1970). Characteristics of repair synthesis in X-irradiated P388 lymphoma cells. *Intern. J. Radiation Biol.*, **18,** 101.

Ayad, S. R., M. Fox and B. W. Fox (1969). Non-semiconservative incorporation of labelled 5-bromo-2'deoxyuridine in lymphoma cells treated with low doses of methyl methanesulphonate. *Mutation Res.*, **8,** 639.

Ayad, S. R., G. R. Barker and A. Jacob (1966). The fractionation of transforming DNA from *Bacillus subtilis. Biochem. J.*, **98,** 3P.

Ayad, S. R., G. R. Barker and J. Weigold (1968). Fractionation of native and denatured transforming Deoxyribonucleic acid from *Bacillus subtilis. Biochem. J.*, **107,** 387.

Ayad, S. R., R. W. Bonsall and S. Hunt (1968). A simple method for production of accurate linear gradient. *Anal. Biochem.*, **22,** 533.

Ayad, S. R., P. Premachandra and R. P. F. Gregory (1971). Fractionation of rapidly-labelled nucleic acids from *Rhodospirillum rubrum* using polylysine kieselguhr column chromatography. *J. Chromatog.*, **58,** 235.

Baguley, B. C., P. L. Bergquist and R. K. Ralph (1965). Fractionation of amino acid acceptor ribonucleic acid on diethylaminoethylcellulose column. *Biochim. Biophys. Acta*, **95,** 510.

Baguley, B. C., P. L. Bergquist and R. K. Ralph (1967). Low molecular-weight T_4 phage specific RNA. *Biochim. Biophys. Acta*, **138,** 51.

Balassa, R. (1955). *In vitro* induzierete transformationen bei Rhizobien. *Naturwissenschaften*, **42,** 422.

Barnhart, B. J., and R. M. Herriot (1963). Penetration of DNA into *H. influenzae. Biochim. Biophys. Acta*, **76,** 25.

Bartoli, F., and C. Rossi (1967). Separation of DNA and RNA by gel filtration. *J. Chromatog.*, **28,** 30.

Bauer, W., and J. Vinograd (1968). The interaction of closed circular DNA with intercalative dye. *J. Mol. Biol.*, **33,** 141.

Bauer, W., and J. Vinograd (1970a). The interaction of closed circular DNA with intercalative dyes. Dependence of the buoyant density upon superhelix density and base composition. *J. Mol. Biol.*, **54,** 281.

Bauer, W., and J. Vinograd (1970b). Interaction of closed circular DNA with intercalative dyes. The free energy of superhelix formation in SV40 DNA. *J. Mol. Biol.*, **47,** 419.

Bautz, E. K. F. (1963). Physical properties of messenger RNA of bacteriophage T_4. *Proc. Nat. Acad. Sci., USA*, **49,** 68.

Bautz, E. K. F., and B. D. Hall (1962). The isolation of T_4-specific RNA on a DNA-cellulose column. *Proc. Nat. Acad. Sci., USA*, **46,** 1585.

Bazaral, M., and D. R. Helinski (1968a). Circular DNA forms of colicinogenic factors E_1, E_2 and E_3 from *Escherichia coli. J. Mol. Biol.*, **36,** 185.

Bazaral, M., and D. R. Helinski (1968b). Characterization of multiple circular DNA forms of colicinogenic factor E_1 from *Proteus mirabilis. Biochemistry*, **7,** 3513.

Bazaral, M., and D. R. Helinski (1970). Replication of a bacterial plasmid and an episome in *Escherichia coli. Biochemistry*, **9,** 399.

Bell, D., R. V. Tomlinson and G. M. Tener (1963). The nucleotide sequences adjacent to the 5′-termini of yeast soluble ribonucleic acids. *Biochim. Biophys. Acta*, **10,** 304.

Bangtsson, S., and L. Philipson (1964). Chromatography of animal viruses on pearl-condensed agar. *Biochim. Biophys. Acta*, **79,** 399.

Bennett, T. P., J. Goldstein and F. Lipmann (1963). Coding properties of *E. coli* leucyl RNA's charged with homologous or yeast activating enzymes. *Proc. Nat. Acad. Sci., USA*, **49,** 850.

Bergquist, P. L., B. C. Baguley, J. M. Robertson and R. K. Ralph (1965). Fractionation of amino acid acceptor ribonucleic acid. *Biochim. Biophys. Acta*, **108,** 531.

Bernardi, G. (1961). Chromatography of native deoxyribonucleic acid on calcium phosphate. *Biochem. Biophys. Res. Commun.*, **6,** 54.

Bernardi, G. (1962). Chromatography of denatured deoxyribonucleic acid on calcium phosphate. *Biochem. J.*, **83,** 32P.

Bernardi, G. (1969a). Chromatography of nucleic acids on hydroxyapatite. I- Chromatography of native DNA. *Biochim. Biophys. Acta*, **174,** 423.

Bernardi, G. (1969b). Chromatography of nucleic acids on hydroxyapatite. II- Chromatography of denatured DNA. *Biochim. Biophys. Acta*, **174,** 435.

Bernardi, G. (1969c). Chromatography of nucleic acids on hydroxyapatite. III- Chromatography of RNA and polynucleotide. *Biochim. Biophys. Acta*, **174,** 449.

Bernardi, G., and S. N. Timasheff (1961). Chromatography of *Ehrlich* ascites tumor cell high molecular weight ribonucleic acid on calcium phosphate. *Biochem. Biophys. Res. Commun.*, **6,** 58.

Bernardi, G., F. Carnevali, A. Nicolaieff, G. Piperno and G. Tecce (1968). Separation and characterization of a satellite DNA from a yeast cytoplasmic 'petite' mutant. *J. Mol. Biol.*, **37,** 493.

Birnsteil, M. L., J. Spieres, I. Purdon, K. Jones and U. E. Loening (1968). Properties and composition of the isolated ribosomal DNA satellite of *X. laevis*. *Nature*, **219,** 454.

Bishop, D. H. L., J. R. Claybrook and S. Spiegelman (1967). Electrophoretic separation of viral nucleic acids on polyacrylamide gels. *J. Mol. Biol.*, **26,** 373.

Bleeken, S., G. Strohbach and E. Sarfert (1966). Estimation of molecular weight of *Escherichia coli* DNA. *Allgem. Mikrobiol.*, **6,** 121.

Bode, V. C., and L. A. MacHattie (1968). Electron microscopy of superhelical circular λ DNA. *J. Mol. Biol.*, **32,** 673.

Bodmer, W. (1965). Recombination and integration in *B. subtilis* transformation: Involvement of DNA synthesis. *J. Mol. Biol.*, **14,** 534.

Bodmer, W. F., and A. T. Ganesan (1964). Biochemical and genetic studies of integration and recombination in *Bacillus subtilis* transformation. *Genetics*, **50,** 717.

Bolton, E. T., and B. J. McCarthy (1962). A general method for the isolation of RNA complementary to DNA. *Proc. Nat. Acad. Sci., USA*, **48,** 1390.

Bolton, E. T., and B. J. McCarthy (1963). An approach to the measurement of genetic relatedness among organisms. *Proc. Nat. Acad. Sci., USA*, **50,** 156.

Boman, H. G., and S. Hjerten (1962). "Molecular sieving" of bacterial RNA. *Arch. Biochem. Biophys.*, Suppl. **1,** 276.

Bourgaux, P., and D. Bourgaux-Ramoisy (1967). Chromatographic separation of the various forms of polyoma virus DNA. *J. Gen. Virology*, **1,** 323.

Bourgaux-Ramoisy, D., N. van Tieghem and P. Bourgaux (1967). Fractionation of polyoma virus DNA on hydroxyapatite: Dependence on tertiary structure. *J. Gen. Virology*, **1,** 589.

Brenner, S., and R. W. Horne (1959). A negative staining method for high resolution electron microscopy of viruses. *Biochim. Biophys. Acta*, **34,** 103.

Bromley, P. A., and R. D. Barry (1970). Fractionation of chick embryo RNA using polylysine kieselguhr column. Personal communication.

Brown, G. L. (1963). Preparation, fractionation and properties of RNA. *Progr. Nucl. Acid Res.*, **2,** 259.

Brownlee, G. G., F. Sanger and B. G. Barrell (1967). Nucleotide sequence of 5S rRNA from *E. coli. Nature*, **215,** 735.

Brownlee, G. G., F. Sanger and B. G. Barrell (1968). The sequence of 5S rRNA. *J. Mol. Biol.*, **34,** 379.

Burton, K. (1956). A study of the conditions and mechanism of the diphenylamine reaction for the colorimetric estimation of deoxyribonucleic acid. *Biochem. J.*, **62,** 315.

Cairns, J. (1961). An estimate of the length of the DNA molecule of T_2 bacteriophage by autoradiography. *J. Mol. Biol.*, **3,** 756.

Cairns, J. (1963). The chromosome of *E. coli. Cold Spring Harb. Symp. Quant. Biol.*, **28,** 43.

Cannon, M. C., and L. K. Dunican (1970). The separation of plasmid and chromosomal DNA from *Staphylococcus aureus* using poly-L-lysine kieselguhr columns. *Biochem. Biophys. Res. Commun.*, **39,** 423.

Caro, L. (1965). The molecular weight of lambda DNA. *Virology*, **25,** 226.

Caspersson, T. (1941). Studien über den Eiweissumsatz der Zelle. *Naturwissenschaften*, **29,** 33.

Chapeville, F., F. Lipmann, G. von Erenstein, B. Weisblum, W. J. Roy, Jr. and S. Benzer (1962). On the role of s-RNA in coding for amino acids. *Proc. Nat. Acad. Sci., USA*, **48,** 1086.

Chen, D., S. Sarid and E. Katchalski (1968). Studies on the nature of messenger RNA in germinating wheat embryos. *Proc. Nat. Acad. Sci., USA*, **60,** 902.

Cheng, T. Y., and N. Sueoka (1962). Heterogeneity of DNA in density and base composition. *Science*, **141,** 1194.

Cherayil, J. D., and H. M. Bock (1965). A column chromatographic procedure for the fractionation of s-RNA. *Biochemistry*, **4,** 1174.

Chervanka, C. H. (1969). *A Manual of Methods for the Analytical Ultracentrifuge.* Published by Spinco division of Beckman Instruments, Inc., Palo Alto, California.

Chevallier, M. R., and G. Bernardi (1965). Transformation by heat denatured deoxyribonucleic acid. *J. Mol. Biol.*, **11,** 658.

Chevallier, M. R., and G. Bernardi (1968). Residual transforming activity of denatured *Haemophilus influenzae* DNA. *J. Mol. Biol.*, **32,** 437.

Chipchase, M. J. H., and M. L. Birnstiel (1963). On the nature of nuclear RNA. *Proc. Nat. Acad. Sci., USA*, **50,** 1101.

Clark, B. F., S. K. Dube and K. A. Marcker (1968). Specific codon-anticodon interaction of an initiator t-RNA fragment. *Nature*, **219,** 484.

Cleaver, J. E., and R. B. Painter (1968). Evidence for repair replication of HeLa cell DNA damaged by ultraviolet light. *Biochim. Biophys. Acta*, **161,** 552.

Cocito, C., A. Prinze and P. DeSomer (1961). Chromatographic analysis of infectious RNA from Poliovirus. *Nature*, **191,** 573.

Cohen, S. S. (1947). The synthesis of bacterial viruses in infected cells. *Cold Spring Harbor Symp. Quant. Biol.*, **12,** 35.

Cohen, S. N., and C. A. Miller (1969). Multiple molecular species of circular R-factor DNA isolated from *Escherichia coli. Nature*, **224,** 1273.

Comb, D. G., and E. Sarker (1967). The binding of 5S ribosomal RNA to ribosomal subunits. *J. Mol. Biol.*, **25,** 317.

Crawford, L. V., and M. J. Waring (1967). Supercoiling of polyoma virus DNA measured by its interaction with ethidium bromide. *J. Mol. Biol.*, **25,** 23.

Crick, F. H. C. (1957). In "DNA and protein synthesis". *Biochem. Soc. Symp., Cambridge, Engl.*, **14,** 25.

Davern, C. I. (1966). Isolation of the DNA of the *E. coli* chromosome in one piece. *Proc. Nat. Acad. Sci., USA*, **55,** 792.

Davis, B. J. (1964). Disc electrophoresis. Method and application to human serum proteins. *Ann. N.Y. Acad Sci.*, **121,** 404.

Davison, P. F., D. Freifelder, R. Hede and C. Levinthal (1961). The structural unity of DNA of T_2 bacteriophage. *Proc. Nat. Acad. Sci., USA*, **47,** 1123.

Detlefsen, M. (1970). Ph.D. thesis, Manchester.

Doctor, B. P., and C. M. Connelly (1961). Separation of yeast amino acid-acceptor ribonucleic acids by countercurrent distribution in modified Kirby's system. *Biochem. Biophys. Res. Commun.*, **6,** 201.

Doctor, B. P., J. Apgar and R. W. Holley (1961). Fractionation of yeast amino acid-acceptor ribonucleic acid by countercurrent distribution. *J. Biol. Chem.*, **236,** 1117.

Doi, R. H., and R. T. Igarashi (1960). Heterogeneity of the conserved ribosomal ribonucleic acid sequences of *Bacillus subtilis. J. Bacteriol.*, **92,** 88.

Doty, P., J. Marmur, J. Eigner and C. Schildkraut (1960). Strand separation and specific recombination in deoxyribonucleic acids: physical chemical studies. *Proc. Nat. Acad. Sci., USA*, **46,** 461.

Dove, W. F., and N. Davidson (1962). Cation effects on the denaturation of DNA. *J. Mol. Biol.*, **5,** 467.

Dyer, T. A. (1967). The nucleic acids of photosynthetic cells. *Phytochemistry*, **6,** 457.

Ellem, K. A. O. (1966). Some properties of mammalian DNA-like RNA isolated by chromatography on methylated bovine serum albumin-kieselguhr columns. *J. Mol. Biol.*, **20,** 283.

Ellem, K. A. O. (1967). A dual-label technique for comparing the rates of synthesis of nucleic acid fractions separated by methylated albumin-kieselguhr chromatography from cells in different states of activity. *Biochim. Biophys. Acta*, **149,** 74.

Erickson, R. J., and W. Braun (1968). Apparent dependence of transformation on the stage of DNA replication of recipient cells. *Bacteriol. Rev.*, **32,** 291.

Erickson, R. L., and J. A. Gordon (1966). Replication of bacteriophage RNA: Purification of the replicative intermediate by agarose column chromatography. *Biochem. Biophys. Res. Commun.*, **23,** 422.

Falkow, S., R. V. Citarella, J. A. Wohlhieter and J. Watanabe (1966). The molecular nature of R-factors. *J. Mol. Biol.*, **17**, 102.

Falkow, S., J. A. Wohlhieter, R. V. Citarella and L. S. Baron (1964). Transfer of episomic elements to *Proteus*. 2—Nature of lac+ *Proteus* strains isolated from clinical specimens. *J. Bacteriol.*, **88**, 1598.

Faulhaber, I., and G. Bernardi (1967). Chromatography of calf thymus nucleoprotein on hydroxyapatite columns. *Biochim. Biophys. Acta*, **140**, 561.

Forget, B. G., and S. M. Weissman (1969). The nucleotide sequence of ribosomal 5S RNA from KB cells. *J. Biol. Chem.*, **224**, 3148.

Fox, M. S. (1960). Fate of transforming DNA following fixation by transformable bacteria. *Nature*, **187**, 1004.

Fraenkel-Conrat, H., and H. S. Olcott (1945). Esterification of proteins with alcohols of low molecular weight. *J. Biol. Chem.*, **161**, 259.

Galibert, F., J. C. Lelong, C. J. Larsen and M. Boiron (1967). Position of 5S RNA among cellular RNA's. *Biochim. Biophys. Acta*, **142**, 89.

Giacomoni, D., and S. Spiegelman (1962). Origin and biological individuality of the genetic dictionary. *Science*, **138**, 1328.

Gillam, I., D. Blew, R. C. Warrington, M. von Tigerstrom and G. M. Tener (1968). A general procedure for the isolation of specific ribonucleic acid. *Biochemistry*, **7**, 3459.

Gillam, I., S. Millward, D. Blew, M. von Tigerstrom, E. Wimmer and G. M. Tener (1967). The separation of soluble ribonucleic acids on benzoylated diethylaminoethylcellulose. *Biochemistry*, **6**, 3043.

Gillespie, D. (1968). Formation and detection of DNA-RNA hybrids. In: L. Grossman and K. Moldavek (Eds.), *Methods of Enzymology XII B*. Academic Press, New York, p. 641.

Gillespie, D., and S. Spiegelman (1965). A quantitative assay for DNA-RNA hybrids with DNA immobilized on a membrane. *J. Mol. Biol.*, **12**, 829.

Glitz, D. G. (1968). The nucleotide sequence at the 3′-linked end of bacteriophage MS2 ribonucleic acid. *Biochemistry*, **7**, 927.

Goldberg, I. D., D. D. Gwinn and C. D. Thorne (1966). Interspecies transformation between *B. subtilis* and *B. licheniformis*. *Biochem. Biophys. Res. Commun.*, **23**, 543.

Goldstein, J., T. P. Bennett and L. C. Craig (1964). Countercurrent distribution studies of *E. coli* B tRNA. *Proc. Nat. Acad. Sci., USA*, **51**, 119.

Goldthwait, D. A., and D. S. Kerr (1962). Studies on the chromatographic properties of amino acid transfer and ribosomal RNA. *Biochim. Biophys. Acta*, **61**, 930.

Goldthwait, D. A., and J. L. Starr (1960). Chromatographic characterization of amino acid transfer ribonucleic acids isolated from yeast. *J. Biol. Chem.*, **235**, 2025.

Goodman, H., and A. Rich (1962). Formation of a DNA-tRNA hybrid and its relation to the origin, evolution, and degeneracy of tRNA. *Proc. Nat. Acad. Sci., USA*, **48**, 2101.

Gould, H. (1966). The specific cleavage of ribonucleic acid from reticulocyte ribosomal subunits. *Biochemistry*, **5**, 1103.

Gross, M., B. Skoczylas and W. Turski (1965). Separation of phenol and ribonucleic acid by dextran gel filtration (Sephadex G-25). *Anal. Biochem.*, **11**, 10.

Grossbach, U., and I. B. Weinstein (1968). Separation of ribonucleic acids by polyacrylamide gel electrophoresis. *Anal. Biochem.*, **22,** 311.

Hall, R. N. (1965). A general procedure for the isolation of minor nucleosides from RNA hydrolysates. *Biochemistry*, **4,** 661.

Hall, B. D., and S. Spiegelman (1961). Sequence complementarity of T_2-DNA and T_2-specific RNA. *Proc. Nat. Acad. Sci., USA*, **47,** 137.

Haruna, I., and S. Spiegelman (1965). Recognition of size and sequence by an RNA replicase. *Proc. Nat. Acad. Sci., USA*, **54,** 1189.

Hayashi, M., M. N. Hayashi and S. Spiegelman (1966). Restriction of *in vivo* genetic transcription to one of the complementary strands of DNA. *Proc. Nat. Acad. Sci., USA*, **50,** 664.

Hayes, W. (1968). *The Genetics of Bacteria and their Viruses*, 2nd ed., Blackwell Scientific Publications, Oxford.

Hershey, A. D., and E. Burgi (1960). Molecular homogeneity of the deoxyribonucleic acid of phage T_2. *J. Mol. Biol.*, **2,** 143.

Higuchi, S., and M. Tsuboi (1966). Interaction of poly-L-lysine with nucleic acids. Poly (A + U), poly (A + 2U) and rice dwarf virus RNA. *Biopolymers*, **4,** 837.

Hinilica, L. S. (1967). Protein of the cell nucleus. *Progr. Nucl. Acid Res*,. **7,** 25.

Hirschman, S. Z., M. Leng and G. Felsenfeld (1967). Interaction of spermine and DNA. *Biopolymers*, **5,** 227.

Hjerten, S. (1962). A new method for preparation of agarose for gel electrophoresis. *Biochim. Biophys. Acta*, **62,** 445.

Hjerten, S. (1964). The preparation of agarose spheres for chromatography of molecules and particles. *Biochim. Biophys. Acta*, **79,** 393.

Hohn, T., and H. Schaller (1967). Separation and chain-length determination of polynucleotides by gel filtration. *Biochim. Biophys. Acta*, **138,** 466.

Holley, R. W., J. Apgar, B. P. Doctor, J. Farrow, M. A. Martini and S. H. Merrill (1961). A simplified procedure for the separation of tyrosine- and valine-acceptor fractions of yeast "soluble ribonucleic acid". *J. Biol. Chem.*, **236,** 200.

Holley, R. W., J. Apgar, G. A. Everett, J. T. Madison, S. H. Merrill and A. Zamir (1963). Chemistry of amino acids specific ribonucleic acids. *Cold Spring Harbor Symp. Quant. Biol.*, **28,** 117.

Holley, R. W., J. Apgar, G. A. Everett, J. T. Madison, M. Marquisee, S. H. Merrill, J. R. Penswick and A. Zamir (1965). Structure of ribonucleic acid. *Science*, **147,** 1462.

Hotchkiss, R. D. (1957). Criteria for quantitative genetic transformation in bacteria. In: W. D. McElory and B. Glass (Eds.), *The Chemical Basis of Heredity*, Johns Hopkins Press, Baltimore.

Huang, R. C. C., J. Bonner and K. Murray (1964). Physical and biological properties of soluble nucleohistones. *J. Mol. Biol.*, **8,** 54.

Hudson, B., and J. Vinograd (1967). Catenated circular DNA molecules in HeLa cell mitochondria. *Nature*, **216,** 647.

Huxley, H. E., and G. Zubay (1960). Electron microscope observation on the structure of microsomal particles from *Escherichia coli*. *J. Mol. Biol.*, **2,** 10.

Ifft, J. B., D. H. Voet and J. Vinograd (1961). The determination of density distributions and density gradients in binary solutions at equilibrium in the ultracentrifuge. *J. Phys. Chem.*, **65,** 1138.

Ikeda, H., and J. I. Tomizawa, (1968). Prophage Pl, and extrachromosomal replication unit. *Cold Spring Harbor Symp. Quant. Biol.*, **33,** 791.

Inman, R. B. (1964). Transitions of DNA homopolymers. *J. Mol. Biol.*, **9,** 624.

Inman, R. B., and R. L. Baldwin (1964). Helix-random coil transition in DNA homopolymer. *J. Mol. Biol.*, **8,** 452.

Ishihama, A., N. Mizuno, M. Takai, E. Otaka and S. Osawa (1962). Molecular and metabolic properties of messenger RNA from normal and T_2-infected *Escherichia coli. J. Mol. Biol.*, **5,** 251.

Iwabuchi, H., H. Kono, R. Oumi and S. Osawa (1965). The RNA components in ribonucleoprotein particles occurring during the course of ribosome formation in *E. coli. Biochim. Biophys. Acta*, **108,** 211.

Jacob, F., and J. Monod (1961). Genetic regulatory mechanisms in the synthesis of proteins. *J. Mol. Biol.*, **3,** 318.

Jaenisch, R., E. Jacob and J. Hofschneider (1970). Replication of the small coliphage M13 : Evidence for long-living M13 specific messenger RNA. *Nature*, **227,** 59.

Javor, G. T., and A. Tomasz (1968). An autoradiographic study of genetic transformation. *Proc. Nat. Acad. Sci., USA*, **60,** 1216.

Jenkins, W. T. (1962). Hydroxyapatite for protein chromatography. In M. H. Coon (Ed.), *Biochemical Preparations*, Vol. 9, John Wiley and Sons, New York, p. 83.

Kaempfer, R., and M. Meselson (1968). Permanent association of 5S RNA molecules with 50S ribosomal subunits in growing bacteria. *J. Mol. Biol.* **34,** 703.

Kano-Sueoka, T., and N. Sueoka (1966). Modification of Leucyl s-RNA after bacteriophage infection. *J. Mol. Biol.*, **20,** 183.

Karau, W., and H. G. Zachau (1964). Isolierung von serinspezifischen transfer ribonucleinsäure fraktionen. *Biochim. Biophys. Acta*, **91,** 549.

Kawade, Y., T. Okamoto and Y. Yamomoto (1963). Fractionation of soluble RNA by chromatography on DEAE ion exchangers. *Biochem. Biophys. Res. Commun.*, **10,** 200.

Kellenberger, E., A. Bolle, E. Boy DeLá Tour, R. H. Epstein, N. C. Franklin, N. K. Jerne, A. Reale-Scafatie, J. Sechaud, I. Bendet, D. Goldstein and N. R. Lauffer (1965). Function and properties related to the tail fibres of bacteriophage T_2. *Virology*, **26,** 419.

Kelly, M. S. (1967). Physical and mapping properties of distant linkages between genetic markers in transformation of *Bacillus subtilis. Mol. Gen. Genetics*, **99,** 333.

Kelmers, A. D. (1966). Preparation of a highly purified phenylalanine transfer ribonucleic acid. *J. Biol. Chem.*, **241,** 3540.

Kelmers, A. D., G. D. Novelli and M. P. Stulberg (1965). Separation of transfer ribonucleic acids by reverse phase chromatography. *J. Biol. Chem.*, **240,** 3979.

Kemp, C. J. (1968). Ph.D. thesis, Manchester.

Kidson, C. (1968). Analysis of DNA fine structure in relation to replication and transcription. *Cold Spring Harbor Symp. Quant. Biol.*, **33,** 179.

Kidson, C. (1969). Analysis of deoxyribonucleic acid structure by partition chromatography. *Biochemistry*, **8,** 4376.

Kidson, C. and K. S. Kirby (1963). Fractionation of DNA. Countercurrent distribution of native and degraded DNA. *Biochim. Biophys. Acta.*, **76,** 624.

Kidson, C., and K. S. Kirby (1964). Fractionation of DNA. Partial strand separation as a basis of fractionation by countercurrent distribution. *Biochim. Biophys. Acta*, **91,** 627.

Kirby, K. S. (1956). A new method for the isolation of ribonucleic acids from mammalian cells. *Biochem. J.*, **64,** 405.

Kirby, K. S. (1962a). Countercurrent distribution of ribonucleic acids. *Biochim. Biophys. Acta*, **61,** 506.

Kirby, K. S. (1962b). Ribonucleic acids. Improved preparation of rat liver ribonucleic acids. *Biochim. Biophys. Acta*, **55,** 545.

Kirby, K. S., J. R. B. Hastings and M. A. O'Sullivan (1962). Countercurrent distribution of ribonucleic acids. *Biochim. Biophys. Acta*, **61,** 978.

Kit, S. (1960). Fractionation of deoxyribonucleic acid preparations on substituted cellulose anion exchangers. *Arch. Biochem. Biophys.*, **87,** 318.

Kleinschmidt, A., and R. K. Zahn (1959). Über desoxyribonucleinsäuremolekeln in protein-mischfilmen. *Z. Naturforsch.*, **14b,** 770.

Kleinschmidt, A., A. K. Burton and R. L. Sinsheimer (1963). Electron microscopy of the replicative form of the DNA of bacteriophage Ø–X174. *Science*, **142,** 961.

Kleinschmidt, A., D. Lang and R. K. Zahn (1960). Darstellung molekularer fäden von desoxyribonucleinsäuren. *Naturwissenschaften*, **47,** 16.

Kleinschmidt, A. K., D. Lang, D. Jacherts and R. K. Zahn (1962). Dartellung und längenmessungen des gesamten desoxyribonucleinsäureinhaltes von T_2-bakteriophagen. *Biochim. Biophys. Acta*, **61,** 857.

Klesius, P. H., and V. T. Schuhardt (1968). Use of lysostaphin in the isolation of highly polymerized deoxyribonucleic acid and in the taxonomy of aerobic *Micrococcaceae*. *J.Bacteriol.*, **95,** 739.

Knight, E., Jr, and J. E. Darnell (1967). Distribution of 5S RNA in HeLa cells. *J. Mol. Biol.*, **28,** 491.

Kubinki, H., and G. Koch (1962). On the separation of Poliovirus ribonucleic acid from cellular ribonucleic acids. *Virology*, **17,** 219.

Kubinki, H., and G. Koch (1963). Instability of an RNA-fraction from Poliovirus infected cells. *J. Mol. Biol.*, **6,** 102.

Kubinski, H., Z. Opera-Kubinska and W. Szybalski (1966). Patterns of interaction between polyribonucleotides and individual DNA strands derived from several vertebrates, bacteria and bacteriophages. *J. Mol. Biol.*, **20,** 313.

Kurland, C. G. (1960). Molecular characterization of RNA from *E. coli* ribosomes. Isolation and molecular weights. *J. Mol. Biol.*, **2,** 83.

Kwan, C. N., D. Apirion and D. Schlessinger (1968). Anaerobiosis-induced changes in an isoleucyl transfer ribonucleic acid and the 50S ribosomes of *Escherichia coli*. *Biochemistry*, **7,** 427.

Lang, D., A. K. Kleinschmidt and R. K. Zahn (1964). Konfiguration und längenverteilung von DNA-molekülen in lösing. *Biochim. Biophys. Acta*, **88,** 142.

Latt, S. A., and H. A. Sober (1967a). Protein-nucleic acid interaction. Oligonucleotide-polyribonucleotide binding studies. *Biochemistry*, **6,** 3293.

Latt, S. A., and H. A. Sober (1967b). Protein-nucleic acid interaction. Cation effect on binding strength and specificity. *Biochemistry*, **6,** 3307.

Lederberg, J. (1955). Recombination mechanisms in bacteria. *J. Cellular Comp. Physiol.*, **45,** (suppl. 2), 75.

Leng, M., and G. Felsenfeld (1966). The preferential interactions of polylysine and polyarginine with specific base sequences in DNA. *Proc. Nat. Acad. Sci., USA*, **56,** 1325.

Lerman, L. S. (1955). Chromatographic fractionation of the transforming principle of the Pneumococcus. *Biochim. Biophys. Acta*, **18,** 132.

Levine, J. S., and N. Strauss, (1965). Lag period characterizing the entry of transforming deoxyribonucleic acid into *Bacillus subtilis*. *J. Bacteriol.*, **89,** 281.

Levinthal, C., A. Keynan and A. Higa (1962). Messenger RNA turnover and protein synthesis in *B. subtilis* inhibited by actinomycin D. *Proc. Nat. Acad. Sci., USA*, **48,** 1631.

Levitt, M. (1969). Detailed molecular model for transfer ribonucleic acid. *Nature*, **224,** 759.

Lewin, B. M. (1970). *The Molecular Basis of Gene Expression*, John Wiley and Sons, New York.

Liquori, A. M., L. Costantino, V. Crescenzi, V. Elia, E. Giglio, R. Pulitt, M. De Santis Savino and V. Vitagliano (1967). Complexes between DNA and polyamines : A molecular model. *J. Mol. Biol.*, **24,** 113.

Loening, U. E. (1967). The fractionation of high molecular weight ribonucleic acid by polyacrylamide gel electrophoresis. *Biochem. J.*, **102,** 251.

Loening, U. E. (1968a). The fractionation of high molecular weight RNA. In: I. Smith (Ed.), *Chromatographic and Electrophoretic Techniques*, Vol. 2, W. Heinemann (Medical books) Ltd.

Loening, U. E. (1968b). Molecular weights of ribosomal RNA in relation to evolution. *J. Mol. Biol.*, **38,** 355.

Loening, U. E. (1969). The determination of the molecular weight of ribonucleic acid by polyacrylamide-gel electrophoresis: The effects of changes in conformation. *Biochem. J.*, **113,** 1313.

Loening, U. E., and J. Ingle (1967). Diversity of RNA components in green plant tissues. *Nature*, **215,** 363.

Lowrie, R. J., and P. L. Bergquist (1968). Transfer ribonucleic acid from *E. coli* treated with 5-fluorouracil. *Biochem.*, **7,** 1761.

Lucy, J. A., and J. A. V. Butler (1955). Fractionation of λ deoxyribonucleoprotein. *Biochim. Biophys. Acta*, **16,** 431.

MacHattie, L. A., and C. A. Thomas, Jr. (1964). DNA from bacteriophage Molecular weight conformation. *Science*, **144,** 1124.

MacHattie, L. A., K. I. Berns and C. A. Thomas, Jr. (1965). Electron microscopy of DNA from *Hemophilus influenzae*. *J. Mol. Biol.*, **11, 648.**

Mahler, H. R., B. Kilne and B. D. Mehrotra (1964). Some observations on the hypochromism of DNA. *J. Mol. Biol.*, **9,** 801.

Mahler, H. R., B. D. Mehrotra and C. W. Sharp (1961). Effects of diamines on the thermal transition of DNA. *Biochem. Biophys. Res. Commun.*, **4,** 79.

Main, R. K., M. J. Wilkins and L. J. Cole (1959). A modified calcium phosphate for column chromatography of polynucleotides and proteins. *J. Amer. Chem. Soc.*, **81,** 6490.

Mandel, M. (1962). The interaction of spermine and native deoxyribonucleic acid. *J. Mol. Biol.*, **5,** 435.

Mandell, J. D., and A. D. Hershey (1960). A fractionating column for analysis of nucleic acids. *Anal. Biochem.*, **1,** 66.

Mangiarotti, G., and D. Schlessinger (1967). Polyribosome metabolism in *E. coli*. Formation and lifetime of mRNA molecules, ribosomal subunit couples and polyribosomes. *J. Mol. Biol.*, **29,** 395.

Marmur, J. (1961). A procedure for the isolation of deoxyribonucleic acid from microorganism. *J. Mol. Biol.*, **3,** 208.

Marmur, J., and D. Lane (1960). Strand separation and specific recombination in deoxyribonucleic acids: biological studies. *Proc. Nat. Acad. Sci., USA*, **46,** 453.

Marmur, J., E. Seaman and J. Levine (1963). Interspecific transformation in *Bacillus*. *J. Bacteriol.*, **85,** 461.

Marmur, J., C. M. Greenspan, K. E. Palace, F. M. Kahan, J. Levine and M. Mandel (1963). Specificity of the complementary RNA formed by *Bacillus subtilis* infected with bacteriophage SP8. *Cold Spring Harbor Symp. Quant. Biol.*, **28,** 191.

Marmur, J., W. F. Anderson, L. Matthews, K. Berns, E. Gajewska, D. Lane and P. Doty (1961). The effects of ultraviolet light on the biological and physical chemical properties of deoxyribonucleic acids. *J. Cellular Comp. Physiol.*, **58,** (Suppl. 1), 33.

Matsuo, K., Y. Mitsui, Y. Ititaka and M. Tsuboi (1968). Interaction of poly L-lysine with nucleic acids. An infrared and X-ray examination. *J. Mol. Biol.*, **38,** 129.

Maxwell, I. H., E. Wimmer and G. M. Tener (1968). The isolation of yeast tyrosine and tryptophan transfer ribonucleic acids. *Biochemistry*, **7,** 2629.

McCallum, M., and P. M. B. Walker (1967). Hydroxyapatite fractionation procedures in the study of the mammalian genome. *Biochem. J.*, **105,** 163.

McCarthy, B. J., and R. B. Church (1970). The specificity of molecular hybridization reactions. *Ann. Rev. Biochem.*, **39,** 131.

McCarthy, B. J., and B. H. Hoyer (1964). Identity of DNA and diversity of mRNA molecules in normal mouse tissues. *Proc. Nat. Acad. Sci., USA*. **52,** 915.

McPhie, P., J. Hounsell and W. B. Gratzer (1966). The specific cleavage of yeast ribosomal ribonucleic acid with nucleases. *Biochemistry*, **5,** 988.

McQuillen, K. (1965). The physical organization of nucleic acid and protein synthesis. In: M. R. Pollock and M. A. Richmond (Eds.), *Function and Structure in Microorganism*, Fifteenth symposium of the society of general microbiology, Cambridge University Press, p. 134.

Meselson, M., and F. W. Stahl (1958). The replication of DNA in *Escherichia Coli*. *Proc. Nat. Acad. Sci., USA*, **44,** 671.

Midgley, J. E. M. (1962). The nucleotide base composition of RNA from several microbial species. *Biochim. Biophys. Acta*, **61,** 513.

Midgley, J. E. M. (1965). Effect of different extraction procedures on molecular characteristics of bacterial ribosomal RNA. *Biochim. Biophys. Acta*, **95,** 232.

Miescher, F. (1871). On the chemical composition of pus cells. Hoppe-Seylers. *Med.-Chem. Untersuchungen*, **4,** 441.

Miyazawa, Y., and C. A. Thomas, Jr. (1965). Nucleotide composition of short segments of DNA molecules. *J. Mol. Biol.* **11,** 223.

Monier, R., S. Naono, D. Hayes, F. Hayes and F. Gros (1962). Studies on the heterogeneity of messenger RNA from *E. coli*. *J. Mol. Biol.*, **5,** 311.

Morell, P., and J. Marmur (1968). Association of 5S RNA to 50S ribosomal subunits of *E. coli* and *B. subtilis*. *Biochemistry*, **7,** 1141.

Muench, K. H., and P. Berg (1966a). Fractionation of tRNA by gradient partition chromatography on sephadex columns. *Biochemistry*, **5,** 970.

Muench, K. H., and P. Berg (1966b). Resolution of aminoacyl transfer ribonucleic acid by hydroxyapatite chromatography. *Biochemistry*, **5,** 982.

Nass, M. M. K. (1969). Mitochondrial DNA. Structure and physical properties of isolated DNA. *J. Mol. Biol.*, **42,** 529.

Nass, M. M. K., and C. A. Buck (1970). Studies on mitochondrial tRNA from animal cells. II. Hybridization of aminoacyl-tRNA from rat liver mitochondria with heavy and light complementary strands of mitochondrial DNA. *J. Mol. Biol.*, **54,** 187.

Nirenberg, M. W., and J. J. Matthaei (1961). The dependence of cell-free protein synthesis in *Escherichia coli* upon naturally occurring or synthetic polynucleotides. *Proc. Nat. Acad. Sci., USA*, **47,** 1588.

Nirenberg, M., P. Leder, M. Bernfield, R. Brimacombe, J. Trupin, F. Rottman and C. O'Neal (1965). RNA codewords and protein synthesis. On the general nature of the RNA code. *Proc. Nat. Acad. Sci., USA*, **53,** 1161.

Nisioka, T., M. Mitani and R. C. Clowes (1969). Composite circular forms of R factor deoxyribonucleic acid molecules. *J. Bacteriol.*, **97,** 376.

Nygaard, A. P., and B. D. Hall (1963). A method for the detection of RNA-DNA complexes. *Biochem. Biophys. Res. Commun.*, **12,** 98.

Nygaard, A. P., and B. D. Hall (1964). Formation and properties of RNA-DNA complexes. *J. Mol. Biol.*, **9,** 125.

Oberg, B., P. A. Albertsson and L. Philipson (1965). Partition studies on nucleic acids. Countercurrent distribution of virus RNA. *Biochim. Biophys. Acta*, **108,** 173.

Oberg, B., S. Bengtsson and L. Philipson (1965). Gel filtration of nucleic acids on pearl-condensed agarose. *Biochem. Biophys. Res. Commun.*, **20,** 36.

Ochoa, S. (1963). Synthetic polynucleotide and the genetic code. *Federation Proc.*, **22,** 62.

Oishi, M., and N. Sueoka (1965). Location of genetic loci of ribosomal RNA of *B. subtilis*. *Proc. Nat. Acad. Sci., USA*, **54,** 483.

Okamoto, T., and Y. Kawade (1963). Fractionation of soluble RNA by a methylated-albumin column of increased capacity. *Biochem. Biophys. Res. Commun.*, **13,** 324.

Olins, D. E., A. L. Olins and P. H. von Hippel (1967). Model nucleoprotein complexes. Studies on interaction of cationic homopolypeptide with DNA. *J. Mol. Biol.*, **24,** 157.

Olins, D. E., A. L. Olins and P. H. von Hippel (1968). On the structure and stability of DNA-protamine and DNA-polypeptide complexes. *J. Mol. Biol.*, **33,** 265.

Ornstein, L. (1964). Disc electrophoresis. Background and theory. *Ann. N.Y. Acad. Sci.*, **121,** 321.

Otaka, E., H. Mitsui and S. Osawa (1962). On the ribonucleic acid synthesized in a cell-free system of *Escherichia coli*. *Proc. Nat. Acad. Sci., USA*, **48,** 425.

Painter, R. B., and J. E. Cleaver (1967). Repair replication in HeLa cells after large doses of X-rays. *Nature*, **216,** 369.

Pakula, R., M. Piechowska, E. Bankowska and W. Walczak (1962). A characteristic of DNA mediated transformation system of two *Streptococcal* strains. *Acta Microbiol. Polon.*, **11**, 205.

Pene, J. J., and W. R. Romig (1964). On the mechanism of genetic recombination in transforming *Bacillus subtilis*. *J. Mol. Biol.*, **9**, 236.

Pettijohn, D. E. (1967). A study of DNA, partially denatured and protein-DNA complexes in the polyethyleneglycol-dextran phase system. *European J. Biochem.*, **3**, 25.

Philipps, G. R. (1969). Primary structure of transfer RNA. *Nature*, **223**, 374.

Philipson, L.(1961). Chromatographic separation and characteristics of nucleic acids from HeLa cells. *J. Gen. Physiol.*, **44**, 899.

Pigott, G. H., and J. E. M. Midgley (1968). Characterization of rapidly labelled RNA in *E. coli*. by DNA-RNA hybridization. *Biochem. J.*, **110**, 251.

Pinck, L., L. Hirth and G. Bernardi (1968). Isolation of replicative RNA from Alfalfa Mosaic virus infected plants by chromatography on hydroxyapatite columns. *Biochem. Biophys. Res. Commun.*, **31**, 481.

Polson, A. (1961). Fractionation of protein mixtures on columns of granulated agar. *Biochem. Biophys. Acta*, **50**, 565.

Porath, J., and P. Flodin (1959). Gel filtration: A method for desalting and group separation. *Nature*, **183**, 1657.

Postel, E. H., and S. H. Goodgal (1966). Uptake of single-stranded DNA in *Haemophilus influenzae* and its ability to transform. *J. Mol. Biol.*, **16**, 317.

Radloff, R., W. Bauer and J. Vinograd (1967). A dye-buoyant-density method for detection and isolation of closed circular duplex DNA: The closed circular DNA in HeLa cells. *Proc. Nat. Acad. Sci., USA*, **57**, 1514.

Ravin, A. W. (1959). Reciprocal capsular transformations of pneumococci. *J. Bacteriol.*, **77**, 296.

Richards, E. G., J. A. Coll and W. B. Gratzer (1956). Disc electrophoresis of ribonucleic acid in polyacrylamide gels. *Anal. Biochem.*, **12**, 452.

Roberts, W. K., and L. D'Ari (1968). Base sequence differences between the ribosomal and "ribosomal precursor" ribonucleic acids from Ehrilch Ascites cells. *Biochemistry*, **7**, 592.

Roberts, J. J., A. R. Crathorn and T. P. Brent (1968). Repair of alkylated DNA in mammalian cells. *Nature*, **218**, 970.

Robins, A. B., and D. M. Taylor (1968). Nuclear uptake of exogenous DNA by mammalian cells. *Nature*, **217**, 1228.

Röschenthaler, R., and P. Fromagenot (1965). Fractionation of *Escherichia coli* s-RNA on dextran gels. (Sephadex). *J. Mol. Biol.*, **11**, 458.

Roger, M. (1964). Fractionation of Pneumococcal DNA following selective heat denaturation: Enrichment of transforming activity for aminopterin resistance. *Proc. Nat. Acad. Sci., USA*, **51**, 189.

Roger, M. (1968). Chromatographic resolution of complementary strands of denatured Pneumococcal DNA. *Proc. Nat. Acad. Sci., USA*, **59**, 200.

Roger, M., and R. D. Hotchkiss (1961). Selective heat inactivity of Pneumococcal transforming deoxyribonucleate. *Proc. Nat. Acad. Sci., USA*, **47**, 653.

Roger, M., C. Beckmann and R. Hotchkiss (1966). The fractionation of denatured Pneumococcal DNA. Evidence for resolution of complementary strands. *J. Mol. Biol.*, **18**, 174.

Rownd, R. (1969). Replication of a bacterial episome under relaxed control. *J. Mol. Biol.*, **44**, 387.

Rownd, R., R. Nakaya and A. Nakamura (1966). A molecular nature of the drug-resistance factors of *Enterobacteriaceae*. *J. Mol. Biol.*, **17**, 376.

Rubenstein, I., C. A. Thomas, Jr. and A. D. Hershey (1961). The molecular weights of T_2 bacteriophage DNA and its first and second breakage products. *Proc. Nat. Acad. Sci., USA*, **47**, 1113.

Rudner, R., J. D. Karkas and E. Chargaff (1968a). Separation of *B. Subtilis* DNA into complementary strands. Biological properties. *Proc. Nat. Acad. Sci., USA*, **60**, 630.

Rudner, R., J. D. Karkas and E. Chargaff (1968b). Separation of *B. Subtilis* DNA into complementary strands. Direct analysis. *Proc. Nat. Acad. Sci., USA*, **60**, 921.

Rudner, R., J. D. Karkas and E. Chargaff (1969). Separation of microbial deoxyribonucleic acids into complementary strands. *Proc. Nat. Acad. Sci., USA*, **63**, 152.

Rush, M. G., and R. C. Warner (1968). Molecular recombination in a circular genome—Ø-X174 and S13. *Cold Spring Harbor Symp.. Quant. Biol.*, **33**, 459.

Rush, M. G., C. N. Gordon, R. P. Novick and R. C. Warner (1969). Circular DNA from *Staphylococcus aureus* and *Shigella dysenteriae* Y6R. *Federation Proc.*, **28**, 532.

Rushizky, G. N., H. A. Sober, C. M. Connelly and B. P. Doctor (1965). Nucleotide sequences in two serine-acceptor RNA components. *Biochem. Biophys. Res. Commun.*, **18**, 489.

Russell, B., T. H. Mead and A. Polson (1964). A method of preparing agarose. *Biochem. Biophys. Acta*, **86**, 169.

Ryter, A. (1967). Relationship between synthesis of the cytoplasmic membrane and nuclear segregation in *Bacillus subtilis*. *Folia. Microbiol., Praglie*, **12**, 283.

Ryter, A. (1968). Association of the nucleus and the membrane of bacteria. A morphological study. *Bacteriol. Rev.*, **32**, 39.

Ryter, A., and F. Jacob (1964). Étude au microscope electronique de la liaison entre noyau et mèsosme chez *Bacillus subtilis*. *Ann. Inst. Pasteur*, **107**, 384.

Saito, H., and Y. Masamune (1964). Fractionation of transforming deoxyribonucleic acid from *Bacillus subtilis* with methylated albumin column. *Biochem. Biophys. Acta*, **91**, 344.

Schachman, H. K. (1959). *Ultracentrifugation in Biochemistry*. Academic Press, New York.

Schaeffer, P. (1958). In The Biological Replication of Macromolecules. *Symp. Soc. Exptl. Biol.*, **12**, 60.

Schildkraut, C. L., and F. Lifson (1965). Dependence of the melting temperature of DNA on salt concentration. *Biopolymers*, **3**, 195.

Schildkraut, C. L., J. Marmur and P. Doty (1962). Determination of the base composition of deoxyribonucleic acid from its buoyant density in CsCl. *J. Mol. Biol.*, **4**, 430.

Schleich, T., and J. Goldstein (1964). Gel filtration properties of CCD-prepared *E. coli*. B s-RNA. *Proc. Nat. Acad. Sci., USA*, **52**, 744.

Schleich, T., and J. Goldstein (1966). Gel filtration heterogeneity of *Escherichia coli*. soluble RNA, *J. Mol. Biol.* **15**, 136.

Schweizer, E., C. Mackechnie and H. O. Halvorson (1969). The redundancy of ribosomal and transfer RNA genes in *Saccharomyces cerevisiae*. *J. Mol. Biol.*, **40**, 261.

Sedat, J. W., R. B. Kelly and R. L. Sinsheimer (1967). Fractionation of nucleic acid on benzoylated-naphthoylated DEAE cellulose. *J. Mol. Biol.*, **26,** 537.

Sedat, J., A. Lyon and R. L. Sinsheimer (1969). Purification of *Escherichia coli* pulse labelled RNA by benzoylated DEAE cellulose chromatography. *J. Mol. Biol.*, **44,** 415.

Semenza, G. (1957). Chromatography of deoxyribonucleic acid on calcium phosphate. *Arkiv. Kemi.*, **11,** 89.

Shapiro, J. T., M. Leng and G. Felsenfeld (1969). Deoxyribonucleic acid-polylysine complexes. Structure and nucleotide specificity. *Biochemistry*, **8,** 3219.

Shepherd, G. R., and D. F. Peterson (1962). Separation of phenol and DNA by sephadex gel filtration. *J. Chromat.*, **9,** 445.

Siegelman, H. W., G. A. Wieczorek and B. C. Turner (1965). Preparation of calcium phosphate for protein chromatography. *Anal. Biochem.*, **13,** 402.

Sinsheimer, R. L. (1959). A single-stranded DNA from bacteriophage Ø–X174. *J. Mol. Biol.*, **1,** 43.

Sinsheimer, R. L., B. Starman, C. Nagler and S. Guthrie (1962). The process of infection with bacteriophage Ø–X174. Evidence for a "replicative form". *J. Mol. Biol.*, **4,** 142.

Slayter, H. S., J. R. Warner, A. Rich and C. E. Hall (1963). The visualization of polyribosomal structure. *J. Mol. Biol.*, **7,** 652.

Smith, I., D. Dubnau, P. Morrell and J. Marmur (1968). Chromosomal location of DNA base sequences complementary to transfer RNA and to 5S, 16S and 23S ribosomal RNA in *B. subtilis. J. Mol. Biol.*, **33,** 123.

Sober, H. A., and E. A. Peterson (1954). Chromatography of proteins on cellulose ion-exchangers. *J. Amer. Chem. Soc.*, **76,** 1711.

Sober, H. A., S. F. Schlossman, A. Yaron, S. A. Latt and G. W. Rushizky (1966). Protein-nucleic acid interaction. Nuclease resistant polylysine-ribonucleic acid complexes. *Biochemistry*, **5,** 3608.

Spahr, P. E. (1962). Amino acid composition of ribosomes from *E. coli. J. Mol. Biol.*, **4,** 395.

Spahr, P. F., and A. Tissières (1959). Nucleotide composition of ribonucleo-protein particles from *E. coli. J. Mol. Biol.*, **1,** 237.

Spitnik, P., R. Lipshitz and E. Chargaff (1955). Studies on nucleoproteins. Deoxyribonucleic acid complexes with basic polyelectrolytes and their fractional extraction. *J. Biol. Chem.*, **215,** 765.

Spizizen, J. (1958). Transformation of biochemically deficient strain of *B. subtilis* by deoxyribonucleate. *Proc. Nat. Acad., Sci. USA*, **44,** 1072.

Stanley, W. M., Jr., and R. M. Bock (1965). Isolation and physical properties of the ribosomal RNA of *E. coli. Biochemistry*, **4,** 1302.

Stephenson, M. L., and P. C. Zamecnik (1962). Isolation of valyl-RNA of a high degree of purity. *Biochem. Biophys. Res. Commun.*, **7,** 91.

Stern, R., and U. Z. Littauer (1968). Fractionation of transfer ribonucleic acid on a methylated albumin-silicic acid column. Preparation of the column. *Biochemistry*, **7,** 3469.

Stewart, C. R. (1969). Physical heterogeneity among *Bacillus subtilis* deoxy-ribonucleic acid molecules carrying particular genetic markers. *J. Bacteriol.*, **98,** 1239.

Studier, F. W. (1965). Sedimentation studies of the size and shape of DNA. *J. Mol. Biol.*, **11,** 373.

Studier, F. W. (1969). Conformational changes of single-stranded DNA. *J. Mol. Biol.*, **41**, 189.

Stuy, J. H., and D. Stern (1964). The kinetics of DNA uptake by *Haemophilus influenzae*. *J. Gen. Microbiol.*, **35**, 391.

Sueoka, N., and T-Y. Cheng. (1962) Fractionation of nucleic acids with the methylated albumin column. *J. Mol. Biol.*, **4**, 161.

Sueoka, N., and T. Yamane (1962). Fractionation of aminoacyl-acceptor RNA on a methylated albumin column. *Proc. Nat. Acad. Sci., USA*, **48**, 1454.

Sueoka, N., J. Marmur and P. Doty (1959). Dependence of the density of DNA on guanine-cytosine content. *Nature*, **183**, 1429.

Summers, W. C., and W. Szybalski (1967). Irradiation of deoxyribonucleic acid in dilute solutions. A sensitive method for detection of single strand breaks in polydisperse DNA samples. *J. Mol. Biol.*, **26**, 107.

Tabor, H. (1961). The stabilization of *Bacillus subtilis* transforming principle by spermine. *Biochem. Biophys. Res. Commun.*, **4**, 228.

Tabor, H. (1962). The protective effect of spermine and other polyamines against heat denaturation of deoxyribonucleic acid. *Biochemistry*, **1**, 496.

Takanami, M. (1967). Nucleotide sequences at the 5—termini of *E. coli.* ribosomal RNA. *J. Mol. Biol.*, **29**, 323.

Tanaka, K., H. H. Richards and G. L. Cantoni (1962). Studies on soluble ribonucleic acid. Partition chromatography of yeast "soluble" ribonucleic acid on sephadex. *Biochem. Biophys. Acta*, **61**, 846.

Thiebe, R., and H. G. Zachau (1965). Zur fraktionierung der löslichen ribonucleinsäure. *Biochem. Biophys. Acta*, **103**, 568.

Thomas, C. A., Jr., T. J. Kelly and M. Rhoades (1968). The intracellular forms of T_7 and P_{22} DNA molecules. *Cold Spring Harbor Symp. Quant. Biol.*, **33**, 417.

Tichy, P., and O. E. Landman (1969). Transformation in quasi spheroplasts of *B. subtilis*. *J. Bacteriol.*, **97**, 42.

Tiselius, A., S. Hjerten and O. Levine (1956). Protein chromatography on calcium phosphate columns. *Arch. Biochem. Biophys.*, **65**, 132.

Tissières, A., D. Schlessinger and F. Gros (1960). Amino acid incorporation into proteins by *Escherichia coli*. K–12 *Proc. Nat. Acad. Sci., USA*, **46**, 91.

Tissières, A., J. D. Watson, D. Schlessinger and B. R. Hollingworth (1959). Ribonucleoprotein particles from *E. coli*. *J. Mol. Biol.*, **1**, 221.

Tomlinson, R. V., and G. M. Tener (1963a). The effect of urea, formamide, and glycols on the secondary binding forces in the ion-exchange chromatography of polynucleotide on DEAE-cellulose. *Biochemistry*, **2**, 697.

Tomlinson, R. V., and G. M. Tener (1963b). A proposed general procedure for isolating end groups of nucleic acids. *Biochemistry*, **2**, 703.

Tsuboi, M., K. Malsuo and P. O. P. Ts'o (1966). Interaction of poly L-lysine and nucleic acids. *J. Mol. Biol.*, **15**, 256.

Vinograd, J., R. Bruner, R. Kent and J. Weigle (1963). Band centrifugation of macromolecules and viruses in self-generating density gradient. *Proc. Nat. Acad. Sci., USA*, **49**, 902.

Volkin, E., and L. Astrachan (1957). In W. D. McElory and B. Glass (Eds.), *The Chemical Basis of Heredity*. Johns Hopkins Press, Baltimore.

Voll, M. J., and S. H. Goodgal (1961). Recombination during transformation in *H. influenzae*. *Proc. Nat. Acad. Sci., USA*, **47**, 505.

Wagner, K. G., and R. Arav (1968). On the interaction of nucleotides with poly L-lysine and poly L-arginine. The influence of nucleotide base on the binding behaviour. *Biochemistry*, **7**, 1771,

Walker, P. M. B., and A. McLaren (1965). Fractionation of mouse deoxyribonucleic acid on hydroxyapatite. *Nature*, **208**, 1175.

Waring, M. (1968). Drugs which affect the structure and function of DNA. *Nature*, **219**, 1320.

Waring, M. (1970). Variation of the supercoils in closed circular DNA by binding of antibiotics and drugs: Evidence for molecular models involving interaction. *J. Mol. Biol.*, **54**, 247.

Weisblum, B., S. Benzer and R. W. Holley (1962). A physical basis for degeneracy in the amino acid code. *Proc. Nat. Acad. Sci., USA*, **48**, 1449.

Weissbach, A., P. Bartle and L. A. Salzman (1968). The structure of replicative Lambda DNA—Electron microscope studies. *Cold Spring Harbor Symp. Quant. Biol.*, **33**, 525.

Williams, D. E., and R. A. Reisfield (1964). Disc. electrophoresis in polyacrylamide gels: Extension to new conditions of pH and buffer. *Ann. N.Y. Acad. Sci.*, **121**, 373.

Wolstenholme, D. R., C. A. Vermeulen and G. Venema (1966). Evidence for the involvement of membranous bodies in the process leading to genetic transformation in *B. subtilis*. *J. Bacteriol.*, **92**, 1111

Wood, D. D., and D. J. L. Luck (1969). Hybridization of mitochondrial ribosomal RNA. *J. Mol. Biol.*, **41**, 211.

Yankofsky, S. A., and S. Spiegelman (1962). The identification of the ribosomal RNA cistron by sequence complementarity. Specificity of complex formation. *Proc. Nat. Acad. Sci., USA*, **48**, 1069.

Yankofsky, S. A., and S. Spiegelman (1963). Distinct cistrons for the two ribosomal RNA components. *Proc. Nat. Acad. Sci., USA*, **49**, 538.

Young, F. E. (1967). Competence in *B. subtilis* transformation system. *Nature*, **213**, 773.

Young, F. E., and J. Spizizen (1963). Incorporation of DNA in the *B. subtilis* transformation system. *J. Bacteriol.*, **86**, 392.

Zachau, H. G., M. Tada, W. B. Lawson and M. Schwieger (1961). Frakionierung de löslichen ribonucleinsäure. *Biochim. Biophys. Acta*. **53**, 221.

Author Index

When there are more than three authors for one paper, they are cited in the text as first author and coworkers. However, the names of the coworkers are given in the references and this index.

Subject Index

An (*) indicates that the subject is indexed as a major heading.